RISK MANAGEMENT IN PROJECT ORGANISATIONS

Peter J. Edwards (MSc, PhD) is an Associate Professor at RMIT University in Melbourne, Australia. Originally trained as a quantity surveyor, he has an extensive background in the construction industry, professionally and academically, and has authored over 90 academic publications. His research and consultancy interests include project risk management and value management, building procurement and sustainable construction.

Paul A. Bowen (BSc (QS), BCom, MSc (Construction Management), PhD) is Professor and Head of the Department of Construction Economics and Management at the University of Cape Town in South Africa. He is a Visiting Professor to RMIT University, a nationally rated researcher and a member of the Academy of Science of South Africa. His professional and academic experience relating to the construction industry has resulted in more than 130 publications. His research interests include project procurement, communication, and the sustainability assessment of buildings.

Peter Edwards and Paul Bowen have enjoyed a professional and academic association spanning more than 25 years, undertaking extensive collaborative research, professional activity and writing together.

RISK MANAGEMENT IN PROJECT ORGANISATIONS

Peter J. Edwards and Paul A. Bowen

ELSEVIER
BUTTERWORTH
HEINEMANN

AMSTERDAM BOSTON HEIDELBERG LONDON NEW YORK OXFORD
PARIS SAN DIEGO SAN FRANCISCO SINGAPORE SYDNEY TOKYO

Published throughout the world excluding Australia and New Zealand by
Butterworth Heinemann
An imprint of Elsevier
Linacre House, Jordan Hill, Oxford OX2 8DP
30 Corporate Drive, Burlington, MA 01803

Published in Australia and New Zealand by
University of New South Wales Press Ltd
University of New south Wales
Sydney NSW 2052
AUSTRALIA

First published 2005

British Library Cataloguing in Publication Data
A catalogue record for this book is available from the British Library

Library of Congress Cataloguing in Publication Data
A catalogue record for this book is available from the Library of Congress

ISBN 075066 629 3

CONTENTS

ACKNOWLEDGMENTS

While the authors willingly accept responsibility for errors and omissions in this text, we also wish to acknowledge the valuable contributions of others. Among these, the contribution of Agustin Chevez Bernaldo de Quiros stands out. He graciously transposed the many diagrams into a format suitable for publishing, and his initiative in establishing a web-based extranet for this book allowed us to easily and rapidly manage the communication of draft material from wherever in the world we found ourselves during the writing process. We are also grateful for the useful comments of many of our colleagues around the world, academic and professional, and for the patient forbearance of our publisher.

We particularly wish to thank our students, past and present, for their willingness to test and explore ideas that were often presented to them in an undeveloped state. Classes and seminars with undergraduate and postgraduate students, at the University of Cape Town, at RMIT University in Melbourne and Singapore, and at the Glasgow Caledonian University in Scotland, have been fertile fields for shaping and refining those ideas. The students' capacity to expand our horizons, beyond the construction industry territory of our own experience, has taken us to the point where we now have the temerity to offer a generic text such as this. In every instance, the learning benefit from these exchanges has been ours far more than theirs. For this reason, we dedicate this book to our students, and indeed to students everywhere, in recognition of their contribution to learning and knowledge.

CHAPTER 1

INTRODUCTION

1.1 AWARENESS ABOUT RISK

Risk is pervasive. It is a universal experience and inescapable. We all face risk – some people more frequently and more willingly than others. While some worry constantly about risk, others cheerfully seek it out. Risk surrounds us, but we are not always fully conscious of it, nor do we consistently respond to it wisely or effectively.

The stories in any daily newspaper will confirm this. Ignore the advertisements, and throughout the paper you will find that government and opposition politicians are making statements about what they perceive as the pressing issues of the day. Editorials and leader articles interpret and comment on these. Under their individual by-lines, foreign correspondents describe conflicts in far-off places. Commentators recall the highlights of yesterday's big sporting contests, repeat gossip about particular players and make predictions about tomorrow's events. On the business pages, the ups and downs of share prices are minutely listed and analysed, and local, national and international corporate news is reported. In the features and entertainment sections, new books, television programs, films and performance arts are reviewed. So, where are the risks in all of this?

If you view this material through the mental lens of a risk perspective, you will perceive risk messages in just about every part of the newspaper. This football player is expected to be in the team line-up for an important match, despite the knee injury he sustained last week – risking further and more lasting damage to the joint. That former pop star is attempting to resurrect her career with a dubious publicity stunt that could backfire. Against all the pundits' sage advice, the government treasurer has announced a budget deficit for the year when a general election is due. A local fire chief believes that, in a fire at a retirement home last week, deaths were averted only because of the warning given by smoke detector alarms installed a few months earlier. The city's lord mayor yesterday farewelled a naval warship carrying troops destined for peacekeeping duties in a foreign trouble-spot, where conflict could flare up again at any time. Road accident statistics for the recent four-day weekend holiday period reached a record high, despite an intensive and expensive publicity campaign to promote safer driving.

These are all stories about risk, either as the consequences of events that have happened or as possibilities for the future. They are also stories about attempts to manage risks. Yet the word 'risk' is rarely mentioned openly. The risk perspectives exist, but are communicated implicitly rather than explicitly.

This implicit treatment of risk is replicated in every area of life. Each of us is constantly surrounded by, and caught up with, circumstances involving risks. Yet none of us consciously spends all our time thinking about and dealing with them. Their sheer prevalence has caused us to develop highly selective attitudes towards of all kinds of risks in our daily lives. Subconsciously, we filter out and ignore those that do not reach some intuitively derived level of seriousness. Or we accept some risks as serious – e.g. the risk of electrocution in the home – but simply assume that adequate protection is already provided in the domestic electrical equipment and devices that we use.

Are such attitudes towards risk reasonable and, since this is a book about project risk management, should the risk awareness situation be any different in a project environment?

1.2 PROJECTS, RISK AND PROJECT MANAGEMENT

An obsessive personal preoccupation with risk is probably unhealthy, but a sufficient awareness of risk (and a capacity to deal with it) is to be encouraged, especially when making decisions that are likely to be life-changing.

The same wisdom applies to projects. Indeed, the nature of modern projects makes the proper management of risks even more important. All projects involve risk, and many of today's projects are inherently more complex than those of yesterday in terms of their structure, technology and resource demands, their financial and organisational arrangements. The objectives set for modern projects have become more demanding. We want them to deliver greater benefits than before – and usually more quickly. Projects are expected to impose fewer negative impacts than their predecessors on sensitive environments. The jargon of projects has expanded to include terms such as 'triple bottom line' (in terms of economic, environmental and social accountability), 'corporate social responsibility', 'due diligence' and 'governance'. All of these are capable of exerting influence on the inescapable decision-making that surrounds projects. They can affect, or are affected by, project outcomes.

All projects have starting points and finishing points. This distinguishes them from other undertakings, such as manufacturing or retail commerce, where the starting point may be known but the finishing point may be theoretically indeterminate.

Some projects are undertaken for the purpose of establishing a facility to house or allow other ongoing activities to take place, for example, a project to install and commission an IT help-desk is undertaken in order to respond to a need to deal with IT-related problems on an ongoing basis. The installation project concludes when the help-desk is operating satisfactorily. Other projects may also incorporate operational and disposal stages following the initial establishment. In events management, for example, a project may include the setting-up activities for an event; the event itself; and the subsequent demobilisation of participants and equipment. Another distinctive feature about projects, therefore, is that they demand the acquisition and application of resources over and above those normally required for purely operational purposes.

Most projects will have objectives that are different from, but related to, the broader objectives of the client or sponsoring organisation. These differences in objectives have implications for project and risk management. Beyond this, projects today tend to involve more participants, partly because of the trend to greater specialisation and partly because the interests of other groups, perhaps ignored in the past, are now increasingly recognised. Each participant has a different role to play, and will have different expectations and needs. The nature and level of the risks that each of these project stakeholders faces will be different.

It can also be deduced that the characteristics of many project risks may themselves change over the life of the project. Project risks thus tend to be dynamic rather than static.

As projects have got more complicated, so too has the way we undertake them. Traditional procedural forms of project procurement have been overtaken by newer and different approaches. This is particularly true for infrastructure mega-projects – dams, power stations, railways, airports, roads – where publicly financed provision and ownership of these types of facilities and services has largely given way to privately financed development, ownership and operational concessionaire arrangements.

All projects must be managed in some way, and all involve the pre-planning and organising of activities that, by definition, will take place at some time in the future. Projects are not created retrospectively in the past. Nor do they occur exclusively in the present: plans cannot be both prepared and executed simultaneously, other than in a virtual environment. Consideration of the future, and making decisions about it, dominate the management processes of all projects. As the future cannot be known with complete certainty, there will be uncertainty associated with any project and, since risk is associated with uncertainty (this is argued further in chapter 2), all projects therefore involve risk. For any project, the extent of uncertainty it exhibits tends to be reflected in the degree of project planning that can be undertaken. Depending upon the particular nature of the project, long-range plans (months, years) have to deal with greater levels of uncertainty than short-range plans (hours, days, weeks). Long-range project plans are therefore more susceptible to risk than short-range plans.

The process of managing projects is itself undergoing development, especially as more and more participants have become involved. In addition to organising and controlling the planning and execution of tasks, the application of technologies, and the acquisition of resources necessary for all projects, modern project management now has to deal with the often competing and conflicting demands of many stakeholders. The tools and techniques of project management have become more sophisticated, providing opportunities for project managers to generate better information and exert greater control over their projects. Advances in information and communication technology (ICT) mean that many of the information requirements of project management can be dealt with more swiftly than ever before, with more controlled distribution. Project extra-nets are an example of this.

However, few of these advances and developments simplify project management. They enable us to manage projects better, but for the most part at the cost of adding more layers of complexity. This issue is discussed further in chapter 3. In the meantime, how do these developments and issues support the need for another book about risk management?

1.3 THE NEED FOR THIS BOOK

Like risk, projects surround us at every point, and in every dimension in life. We are a project-driven society. The growth of project management as a professional discipline is testimony to this, although obviously not all projects are professionally managed. Children are confronted with projects from the very earliest days of formal schooling. Adults in every field of employment are increasingly expected to be involved with them at various times during their careers. Even finding a new job or changing one's career is today regarded as a project. There is no escape for retirees, either, as they take up new interests and new involvements in later life.

Society desires that all projects should be successful, and has become less tolerant of failure. Pressure is exerted on project managers to minimise the chance that failure will happen. The increasing pressures upon projects and their management suggest that it is prudent for anyone involved in a project to be concerned about the risks associated with that project, and about how their risks should best be managed.

There is a significant growth in risk management education and training at tertiary level, and a proliferation of professional career development courses on the same topic. Many of these resources are tailored to the requirements of specific industries and professions. Few are presented in a project context. That they have become available is due more to a spreading concern with risk generally in contemporary society than to any particular appreciation of risk issues in project management. We struggle to understand and deal with many risk situations – the environment, finance, health, violence and terrorism included – from a domestic level through to corporate and government levels; and from a personal to national and international perspectives. Worldwide concern also explains the rise in risk management consultancy, often offered as an adjunct to their existing services by professional firms such as accountants. The risk management advice rarely has a project focus. The emergence of all these factors drives a demand for information about risk and hence a search for suitable books.

There are many books about risk, and an increasing number about risk management, but here again few bring these together under a project banner. None looks at how risk management could be used as a way to improve project performance. This book is based upon the premise that the management of projects can be improved by firstly raising organisational awareness about risks, and then implementing formal processes to deal with them and learn from them. Doing this increases the likelihood that projects will be successful.

1.4 THE APPROACH OF THIS BOOK

The newspaper analogy above provides a clue to the approach adopted in this book. All media are intended to communicate messages that convey meaning to receivers. Risk is only perceived and understood if the relevant messages, in a risk-rich environment, are effectively communicated. This is so whether the risk messages are intra- or inter-personal in nature; that is, whether they are inwardly perceived by individuals or jointly understood between two or more people. I can identify and deal systematically with my personal risks only if I acknowledge and act (intra-personally) upon the risk-related messages generated by my brain, which in turn is responding the stimuli received from other external sources. In truly life-threatening situations, of course, reflexive physiological responses might come into play in an instinctively immediate, untrained and unrehearsed way, but such situations are thankfully rare for most of us. Under most circumstances we have the time and opportunity to take a more deliberative approach to the risks we face and, even for situations of personal danger, pre-event training and simulation can sharpen and improve the effectiveness of our reactions (hence the preparation value obtained from the intensity and frequency of the training and exercises undertaken by sportspeople and military forces).

In group or team situations, we can properly begin to manage our risks only if we share a common understanding about them, and our risk messages are inter-personally (or inter-organisationally) effective. This book therefore explores project risk and risk management from a communication perspective, using communication theory where necessary to compare the *normative* (what should be) with the *descriptive* (what is) risk management practices of project participants.

We also contend that project risks are not really the risks of the project itself, since projects are impersonal, but are those faced by the stakeholders (participants or actors) engaged in the project and acting in some decision-making capacity. Because projects are brought to fruition through the organised effort of at least some of these stakeholders, it will be necessary to consider the organisational structures of projects, and of the project stakeholders.

The content and emphasis of the book is descriptive and discursive, rather than mathematical. It focuses upon describing risks and risk management processes, and upon discussing issues. This focus is deliberate, and arises from our experience in dealing with students and clients. There can be no denying the importance of mathematics in the assessment processes needed for some risks, where differences in the magnitude of the factors that contribute to a particular risk can render it critical to the whole success of the project. On the other hand, the most sophisticated mathematical risk analysis treatment will be useless and wasted if the fundamental nature of that risk is not properly understood, if the risk itself has not been properly identified, or if the risk parameters do not justify such analytical treatment. We will point to instances where we believe the application of mathematical techniques is appropriate, but the book will intentionally dwell more upon concepts of risk and the processes of risk management in projects. In this regard, then, the book takes a qualitative rather than a quantitative approach.

1.5 PROJECTS AS A GENERIC PHENOMENON

The uniqueness/similarity paradox of projects exposes a dilemma for the authors of any book about project management. Should they focus upon the micro-management needs of specific types of projects, or attempt a macro-management approach that concentrates solely upon important common project characteristics?

This book is intended to be a generic text, applicable to all types of projects. Since we have already noted that projects are invariably diverse in terms of type, scope, size, technology and resource requirements, cost, value, duration, location and organisation, this means that some loss of specific focus will inevitably occur at times in the following chapters. Even for projects of similar type, e.g. school construction projects, other factors associated with them, such as the individual site locations, ground conditions, neighbouring buildings, starting and finishing dates, etc., will ensure that each remains unique. Project diversity, therefore, can extend across the full range of task, technology, resource and organisational requirements that all projects entail.

While it is not possible to be specific in this text about particular risks for every type of project, and yet at the same time provide sufficiently comprehensive coverage of the topic, many examples are presented. The intention of the book is to use these examples to illustrate principles and techniques of risk and risk management that are generically applicable. There is no attempt to provide exhaustive lists of risks for particular types of projects. That is the challenge that always faces you, the reader, as you seek to identify and manage the risks of your own unique projects. What you will find instead are some useful tools and techniques to help you in this responsibility.

1.6 THE OBJECTIVES OF THE BOOK

Our aim in this book is to enhance your awareness and understanding of the presence and nature of risk in a project environment. We hope that it will encourage you to consider project risks more carefully, and that it will help you to develop confidence in dealing with the risks associated with your projects in a systematic manner. It should also demonstrate the critical importance of effective communication in project risk management practice.

As a generic project management text, the book is intended to serve the interests and needs of project managers and students in many disciplines and professions including architecture, construction, engineering, property, commerce, health, IT, finance and banking, telecommunications, education, entertainment, events management and public administration. It should be useful to anyone involved in organisational decision-making for projects.

1.7 THE STRUCTURE OF THE BOOK

Primarily, this book is about projects and project organisations, and about risk and risk management.

Chapter 2 establishes a starting point by examining the conceptual nature of risk. Risk and uncertainty are discussed, from theoretical and practical standpoints. A risk classification system is explored as a means of categorising different risks.

Chapters 3 focuses upon the nature of projects; the factors that contribute to project complexity; the risks which projects generate; and issues that tend to make projects 'risky' undertakings.

Organisational theory, decision theory and communication theory dominate chapters 4 and 5, but none of the material presented is intended as a stand-alone exposition on any of those topics. The purpose is to show how contemporary thinking in these fields can and should influence management approaches to dealing with risks.

From this point, the book turns towards more practical matters. Given the preceding background and theoretical frameworks, chapters 6, 7 and 8 propose and describe a systematic approach to risk management. This material is written mainly from a procedural point of view, covering many of the processes and techniques currently available for identifying and analysing risks, for choosing appropriate response treatments, for progressively monitoring and controlling risks during the project life, and for collecting risk knowledge after a project has finished.

Before it can be used, however, an effective risk management system first has to be designed and implemented in an organisation. This is considered in chapter 9.

Chapter 10 concludes with a brief discussion of opportunity management. This is a controversial development in risk management, and is explored here as a logical counter to the more generally held negative view of risk.

The structure and content of the book are intended to suit the immediate needs of its target readership. It is aimed mainly at the perceived curriculum requirements of postgraduate project management students in many disciplines. The authors do not pretend to provide an exhaustive theoretical understanding of risk, nor even of risk management. Contiguous aspects of cognitive and behavioural psychology, for example, have not been covered. Nor does the book present a complete treatment of the theories of organisational structure, decision-making or communication. For these reasons, engineers are likely to find the mathematical treatment of risk lacking. Technically minded people might seek more information about processes. Managers may want to know more about people-related risk issues. A concise text such as this is incapable of dealing with every facet of all of these topics, but hopefully it should whet your appetite. If it encourages you to extend your reading in some or all of the many knowledge areas in the sciences and humanities that affect risk and its management, the book will have more than served its purpose.

CHAPTER 2

WHAT IS RISK?

2.1 INTRODUCTION

This chapter considers definitions and concepts of risk. An argument is made for distinguishing between risk and uncertainty. The classification of risk is explored, and generic categories for risks are proposed.

2.2 DEFINITIONS AND CONCEPTS OF RISK

Definitions of risk

Two joint Australia/New Zealand standards refer to risk. Each offers slightly different definitions. AS/NZS 3931 (1998), which covers risk analysis of technological systems, defines risk as: 'the combination of the frequency, or probability, of occurrence and the consequence of a specified hazardous event', succinctly noting that the concept of risk always incorporates two elements: probability (of occurrence), and consequence. AS/NZS 4360 (1999), which deals with risk management, states that risk is: 'the chance of something happening that will have an impact upon objectives. It is measured in terms of consequences and likelihood'. Here the notion of objectives is introduced. In these definitions, 'probability', 'likelihood' and 'chance' are used synonymously, as also are 'consequence' and 'impact'.

According to the *Oxford English Dictionary* (OED, 1989), risk is: 'the chance or hazard of a commercial loss'. From the same source, two further contextual dictionary definitions are given: 'exposure to mischance or peril' and 'the chance that is accepted in economic enterprise and is considered the source of profit'. A dual view of risk – loss and gain – is introduced here; hence the popularity of 'upside risk' and 'downside risk' in some financial circles.

This dual view, however, has to be set against the fact that few, if any, texts about risk and risk management present examples of risk analysis dealing with purely positive outcomes. Also, society generally seems to have come to prefer the notion that risk is concerned more with negative, than with positive, outcomes.

Despite the largely negative connotation of risk that prevails today, it has to be conceded that one person's risk may be another's opportunity to profit. In later chapters we shall show that, while risk itself may be regarded as a negative phenomenon, approaches to risk management can be positively framed. Later still, we will consider the need for opportunity management.

So far, it appears that risk comprises three components: the probability that an event will occur, the event itself, and the impact or consequences of that event. However, this overlooks an important fourth factor.

A Royal Society (1991) report suggests that: 'Risk is the probability that an adverse event occurs during a stated period of time.' This is a narrower definition in one sense, since 'adverse' excludes the notion of opportunity for risk gain. On the other hand, risk is now constrained by time. This makes sense, given that the probability of occurrence, the impact, or both, of particular risk events may change over time. Indeed, few risks continue unchanged indefinitely, and some disappear completely after a specific time period. This is particularly true for project risks, since we have already noted in the first chapter that projects have starting and finishing points.

Whichever definition is preferred, the important point to remember is that, in dealing with risk, all four aspects should be considered:

- the probability that an event will occur;
- the event and its nature;
- the consequences of that event; and
- the period of exposure to the event (and to its consequences if that is also relevant).

Risk as a social construct

The concept of risk is sociologically framed. We derive our understanding about risks, and our attitudes towards them, largely from the society in which we live and work. A community that knew nothing about money, as a means of payment for goods and services, would not appreciate financial risk (but might instead have a comprehensive awareness of the risks of bartering goods and services). A society that had no knowledge of the practice of human surgery could have no understanding of surgical risk. Similarly, a religious sect with particular values and beliefs might understand the risk of dying through an outbreak of food poisoning by ascribing it to the divine disposition of some higher being or force. For most of the time, this notion of risk as a social construct makes little difference to the way in which we treat it but, if you are working in a cross-cultural environment, the way in which risk is understood in that environment cannot be ignored. For example, the approach to occupational health and safety on a construction project in a country such as Singapore, with its extensive use of foreign labour drawn from many countries, needs to be carefully considered, especially in terms of communicating safe working practices to on-site workers.

Another point is that we do not have to experience a risk personally in order to understand it. Most adults are perfectly capable, for example, of imagining the cause and consequences of a vehicle accident without having to first experience one. Our understanding may not be quantitatively precise, but it is sufficient (or should be) to encourage us to drive more carefully.

Risks are therefore perceived and experienced by *people*, whose understanding of them is influenced by the degree to which they accept the values and beliefs of the society in which they live, and by their ability to assess the capacity of those risks to affect their lives.

This social view of risk is important for project management in two respects. It is people (through the organisations in which they are involved) who experience risks, not projects. It is people who must manage risks (or at least manage the people who will deal with the risks operationally). Effective project risk management should therefore reflect this. We will return to these two points in later chapters. First, however, we must cover a good deal of related background knowledge, and this starts by considering the contexts of risk.

Risk contexts

Risk is contextual. It arises in the context of a situation that exists or is likely to occur at some point in the future.

The situations which most give rise to risks are those which involve us in engaging in activities, carrying out tasks, making commitments or entering into obligations. Thus, if we participate in so-called 'extreme' sports, we face the possibility of being killed or injured. If we offer to wash and dry the dishes after dinner in a friend's home, we may endanger the relationship if precious crockery gets damaged in the process. If we commit ourselves to marriage we face the possibility of subsequent divorce and, if we agree to stand bail for someone charged with a criminal offence, there is a chance that we will have to forfeit a large sum of money should he or she fail to appear in court to answer the charge.

Note that for each of theses contexts, the risks arise out of decisions on our part to participate in the activity, carry out the task, make the commitment, or enter into the obligation. (The risk contexts for projects are explored more fully in chapter 3, while project decision-making is considered in chapter 5.)

Given this brief introduction to the context of risk, it is appropriate to consider two more aspects of the conceptual nature of risk, and their implications. These aspects are the mathematical and behavioural concepts of risk.

Mathematical concepts of risk

Ancient Greek philosophers discussed uncertainty, and Chinese mathematicians explored techniques of probability, at least two thousand years ago. Formal a priori understandings of risk, however, in terms of the associated mathematical probabilities and their treatment, date from the seventeenth-century European mathematicians. Christiaan Huygens (1629–95) explored the concept of mathematical expectation, and wrote a formal treatise on probability in 1657. The large Bernoulli family of mathematicians continued this work. Laplace (1749–1827) pursued an analytical theory of probability and Poisson (1781–1840) explored probability as a necessary part of decision-making (judgment). In the eighteenth century, the parson-mathematician Thomas Bayes proposed his theory concerning the effect of introducing new information on the use of probability to maximise expected return.

As with most applications of mathematics, virtually all of the mathematical exploration of probability relies on the principle of experimental repeatability. Taking the mathematical approach, risk situations can therefore be replicated over and over in a manner similar to physical or chemical experimentation, and the findings analysed with statistical reliability.

While this approach may be possible in industries such as manufacturing, where mechanised and computerised repetitive production-line processes are encountered, it is less frequently applicable to projects. Projects tend to be unique, non-serial undertakings, and it is rare for stakeholders to be able to base their decision-making on repeated prior experimentation. Mathematical simulation (e.g. Monte Carlo simulation, where iterative random number generation is used to represent probabilities in order to select unique variable values from a predetermined range) techniques can be used to replace experiments, but if the validity of the input data is weak, then the outputs of simulation models cannot be robust.

This is not to say that mathematical concepts of risk have no place in project management. It is simply that the nature of projects tends, for the most part, to limit the occasions where they can be properly and reliably exploited.

Behavioural concepts of risk

Compared with its mathematical counterpart, the scientific investigation of risk, in terms of human behaviour and decision-making under uncertainty, has been a much more recent undertaking. Von Neumann and Morganstern (1944) reported early theories of human behaviour under conditions of economic uncertainty. Slovic (1972), Cohen (1979, 1981), and Kahnemann and Tversky (1979, 1982) have been prolific researchers and authors in this field in the twentieth century, and have produced seminal work on human decision-making and judgment under uncertainty, with a focus on the pathology of heuristics and biases. It should be remembered, however, that that these authors are reporting the findings of experiments designed to explore the nature of human judgment about (possible) future events, and not generally concerned with issues relating to the management of risk, nor with post-risk decisions (i.e. what should happen after a risk event has occurred). We will explore the latter more comprehensively in chapter 8 on risk management systems.

The behavioural approach to risk is important in project management, where decision-making is often a matter of choosing between a limited number of alternatives, and where the consequences of the decisions will impact upon the project outcomes and upon the decision-makers.

Risk and decision-making

As noted earlier, risk arises out of individual or organisational decision-making. At one extreme, this might involve whole societies. A nation, for example, through democratic processes of government, might decide to develop and implement a totally new social welfare policy. The decision-making in this case will flow through the myriad parties and procedures involved in such a project: government committees; planning groups; design consultants; project managers; administrators, and many others. It will be informed by consultation with appropriate interest groups, and perhaps even by public meetings and protest lobby action. In every instance, the decision-making will be associated with project activities relating to undertaking tasks or commitments, in order to achieve the project objectives. Each decision-maker in the process is faced with the possibility that events may occur which will affect the fulfilment of these tasks, commitments and objectives.

In a narrower context, the decision-making may be individual. For example, you might decide to embark on a project to boil an egg for breakfast. The decisions flowing from your objective of achieving a nicely soft – neither runny nor hard – four-minute egg will surely expose you to several of the technical risks associated with culinary projects!

Any exploration of project risks therefore needs to consider not only the tasks, commitments and objectives of the project, but also the decision-making processes associated with the organisational structures (the stakeholders and their relationships to each other) through which projects are procured. While we will follow project management convention and refer to 'project risks', this should not disguise the fact that these are actually the risks of the stakeholders in such projects. We will consider the implications of this proposition more fully in later chapters. For now, however, we must reflect upon the association of risk and uncertainty, as the latter is inevitably encountered in most project decision situations.

2.3 RISK AND UNCERTAINTY

A clear distinction between risk and uncertainty is difficult to draw. Hertz and Thomas (1984) suggest that: 'risk means uncertainty and the results of uncertainty ... risk refers to a lack of predictability about problem structure, outcomes or consequences in a decision or planning situation'. Cooper and Chapman (1987) also associate risk and uncertainty: 'Risk is ... as a consequence of the uncertainty associated with pursuing a particular course of action.'

The Australian/New Zealand Standard on technical risk (AS/NZS 3931, 1998) notes that uncertainty is connected with risk, and that it may be encountered in the risk data or in the risk models employed for analysis. The AS/NZS 3931 Standard advises that it is necessary to translate the uncertainties into risk model parameters.

The OED (1989) provides several definitions of uncertainty: 'the quality of being uncertain in respect of duration, continuance, occurrence, etc.'; 'indeterminate as to magnitude or value'; 'state of not being definitely known'; 'the quality of a business risk which cannot be measured and whose outcome cannot be predicted or insured against'.

Bennett and Ormerod (1984), in the context of construction project scheduling, believe that uncertainty is a collection of factors that contribute to construction problems, and is subdivisible into variability in task performance and interference in task progress. They note that uncertainty must be dealt with explicitly and systematically. However, while simulating, or modelling, a risk situation has the benefit of recognising the presence and measuring the effect of uncertainty, the extent of simulation that is possible to undertake in analysing risks in this way is almost always limited by practical considerations.

Mawdesley, Askew and O'Reilly (1997) state that uncertainty is: 'the quality associated with an event which results in an inability to predict its outcome accurately'. Their definition supports the notion of unpredictability, but introduces a further concept of accuracy – which would then lead to the question: accuracy in terms of what? Smith (1999) puts it more simply, by associating uncertainty with the 'unknown', as compared with the 'known', properties of risk.

From all of this, it seems reasonable to accept that risk and uncertainty are in some way associated; and that the association may have to do with particular courses of action, risk data, and predictive risk models (simulations). Logically, uncertainty is some state short of certainty; implying inadequate or incomplete knowledge about the subject at issue. By extension, dealing with uncertainty (e.g. in risk analysis) will require either the acquisition of additional information in order to eliminate or reduce the level of uncertainty, or the adoption of assumptions to substitute for the lack of complete knowledge.

The effort of gathering additional information leads to the practical problem of simulation noted earlier. There will always be a trade-off between the cost of the effort needed to obtain the extra information and incorporate it into a suitable model, and the benefit to be derived from the greater level of certainty achieved. For most project situations, therefore, assumptions have to be made. It is the level of assumptions made which constitutes much of the work of risk analysis, and flows on into the subsequent response and monitoring processes of risk management.

This assumptive understanding of uncertainty can be illustrated diagrammatically. Imagine a situation in which an estimator is investigating the consequences of varying worker productivity upon a company's tender pricing for a project. The estimator has access to case data from previous projects, which will permit the objective formulation of a cumulative distribution function for the workers' historical outputs.

Figure 2.1 shows a frequency (or probability mass) function for the hypothetical historical case data. For the purposes of the example, it does not matter if these are mean production rates taken from different projects or rates observed at random on one or more projects. The estimator has a record of only one case where a worker has produced less than 250 units per day on a project, and only one case of a rate in excess of 800 units per day. Since the workers are not machines, their daily output will not be consistent and is likely to be influenced by a variety of physical, psychological, environmental and operational factors.

While it is theoretically possible for a worker to produce less than 250, or more than 800 units per day, by adopting this frequency

Figure 2.1 Frequency function for worker productivity data

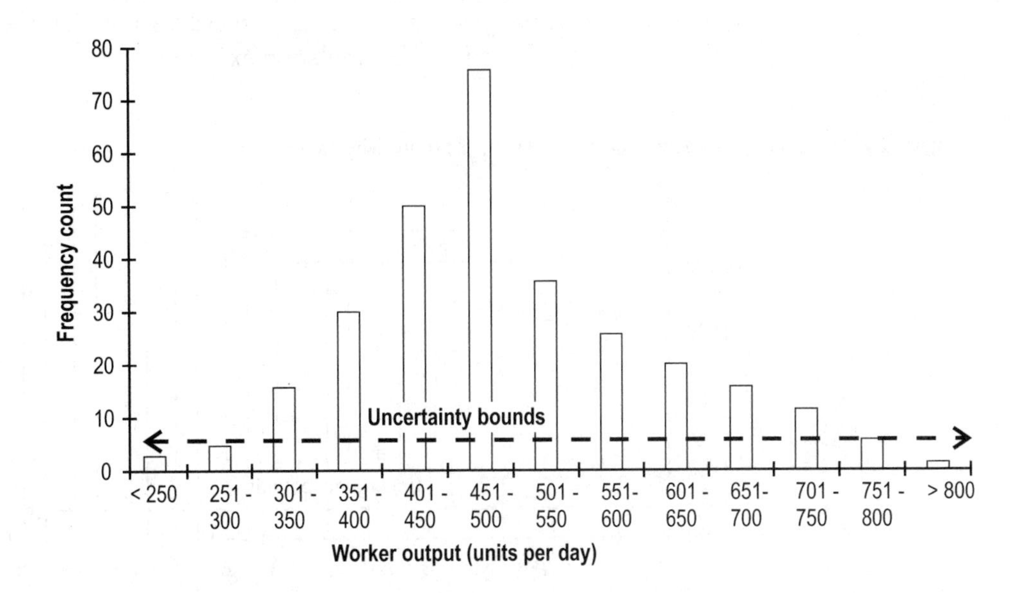

function the estimator has *bounded* the extent of uncertainty associated with the estimating objective. The uncertainty arises because it cannot be known with certainty what production rate will be achieved if the tender is successful and the project goes ahead. By setting the production rate boundaries, the estimator has constrained the uncertainty for the purposes of tender decision-making, and is adopting a frequency approach to probability.

These hypothetical data may be re-arranged to form a cumulative distribution function, with the frequencies transformed to probabilities, as shown in figure 2.2. The same uncertainty boundaries will apply, but now it is possible to explore the probabilities (risks) associated with estimating decisions. Thus, there is a 0.07 (0.89 – 0.82) chance that the actual output rate will lie between 600 and 650 units per day. There is just over a 0.6 chance that the actual rate will be less than 500 units per day (Choice A: the modal value of figure 2.3). Note that the estimator's risk lies in the possibility of the actual output rate turning out to be less than the estimated rate. If the estimator wishes to reduce this risk, Choice B would entail a 0.17 chance that the rate would be less than 400 units per day; but incorporating this more pessimistic rate into the tender would increase the estimated cost of the work and thus increase the risk that the tender would not be sufficiently competitive to win the job. Basing the tender on higher productivity rates reduces the risk of losing the job, but imposes greater pressure on production management to ensure that the workers achieve the output rates required for a profitable project outcome.

Unforeseen factors may also intervene (the *interference* referred to earlier) to affect the actual production rates, but these are not identified and therefore not reflected in the simulation example.

Figure 2.2 Cumulative distribution function for worker productivity data

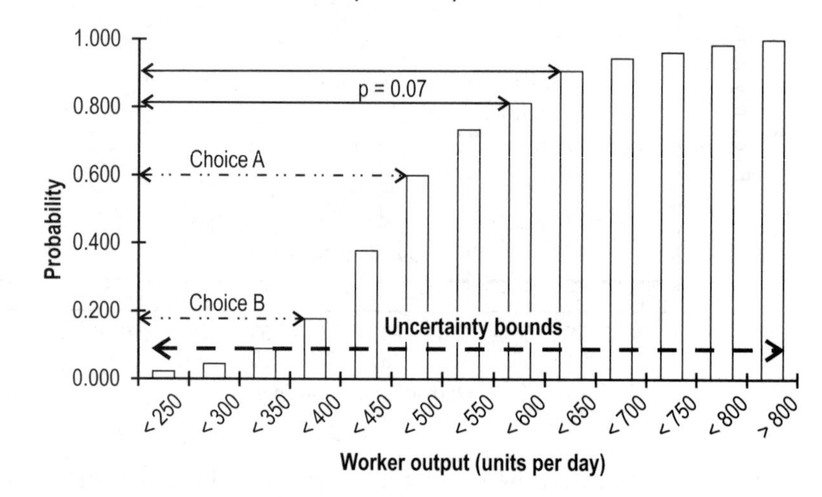

This example shows how uncertainty has to do with particular courses of action (the estimator's output rate decision). The uncertainty is concerned with risk data intended for use in a risk model. We can see how uncertainty arises; how it is associated with risk; and how it is constrained (or bounded) by assumptions for the purposes of decision-making. In this case, the estimator could use Monte Carlo simulation to calculate a 'best estimate' of the predicted productivity rate, and use this rate as the basis for the tender decision, knowing the risk involved.

In most real-life projects, the mass of objective data hypothesised for the project estimating example is rarely available or obtainable, and thus the process of bounding the uncertainty becomes more difficult. It can still be done.

An alternative approach to bounding uncertainty is the adoption of simpler distributions. Typical among these is the three point triangular distribution, where the estimator forecasts the highest and lowest expected values for the productivity rates, together with a 'most likely' rate, in order to establish a basis for producing frequency distribution and cumulative frequency function curves, using Monte Carlo simulation to 'smooth' the curves, and then sampling from these to establish specific risk levels. Instead of relying on one estimator's forecasts alone, it might be possible to use Delphi techniques, by canvassing the views of an expert panel of estimators, to obtain a consensus view of the highest, lowest and 'most likely' values for the distribution. With this technique, each panel member is asked to offer an independent opinion for the three values. The descriptive statistics – range, mean and standard deviation – for the panel members' values are calculated. This information is fed back to the experts, who are now asked if they wish to amend their (independent) opinions. Descriptive statistics for the second 'round' of the panel members' value opinions are calculated, and the process is repeated in subsequent 'rounds' until no further change in the statistics can be detected, and thus the concensus view is gained.

Another approach is possibly even more subjective. Philosophers exploring economic expectation theories have argued that it is rationally defensible to treat the irreducible uncertainty of decisions in economics, not in terms of traditional probabilities, but in terms of the capacity of the outcomes to surprise. Thus, an experienced estimator might be surprised if workers were to achieve a production rate of less than *w* or more than *x* units per day, and highly surprised if their rate was less than *y* or more than *z* units per day. In other words, we can bound the inherent uncertainty by the degree of surprise we are prepared to experience.

All this is really the subjective aspect of risk analysis. The element of human judgment is brought into play, and this judgment is in turn influenced by the intrusion of human biases and the propensity to make errors. However, as we shall see later in looking at risk analysis, it is not always essential to know with precision the mathematical probabilities of occurrence and impact magnitude of each risk in order to manage it. For many risks, approximate values will be sufficient.

Two further examples serve to demonstrate more clearly the association of risk and uncertainty, and the dilemma of dealing with uncertainty.

In the first text box, example 2.1 describes a project developed to help combat the rising incidence of crimes related to domestic violence in the United States. The model, upon which the project was based, is explicitly intended to explore the risk of serious assault for potential victims who have already experienced some form of threat at a lower level. A great deal of uncertainty is obviously involved in trying to predict the likelihood of more serious, or even deadly, assault occurring in the future. The model bounds that uncertainty, for the purposes of prediction, by comparing a victim's subjective assessments of his or her aggressor's behaviour and attitudes with a database of 18 000 assessments culled from the records of actual assault incidents in the past. In other words, the model assumes that the 'most likely' outcome in the present case will not be worse than the worst outcome from the cases in the database which best match the potential victim's subjectively based assessment inputs.

Example 2.1

ASSAULT VICTIM HARM MINIMISATION PROJECT

In the United States, the continuing rise in the number of incidents of domestic violence has been a matter of concern. A project was undertaken to trial a computer-based predictive model that would assess the level of risk for a potential assault victim and provide an indication of the best strategy for the victim to minimise this risk.

Data from over 18 000 recorded incidents were used to build the database for the model. Using the model requires input answers to more than forty questions. The questions, mostly of the dichotomous yes/no or rating types, explore the victim's circumstances and his or her perceptions about the aggressor's actions and attitudes. The answers are elicited through verbal questioning by a skilled interrogator who enters them into the model via a computer keyboard. The model then compares these answers with the characteristics and violent (or non-violent) actions of known offenders from the incident database.

The model output is mainly text-based, and therefore linguistically qualitative. It includes a profile of the alleged aggressor, an indication of the potential seriousness of the situation (i.e. a display of the possible escalating levels of the risk of serious or even deadly assault occurring if the current situation is allowed to continue) and offers alternative strategies for the potential victim to adopt in order to try to reduce or avoid the risk.

Given that we have already referred to the human propensity to make errors, and coupling with this the vagaries of human behaviour; it might seem reasonable to question the reliability and usefulness of the model described in example 2.1. After all, it is based upon data relating to the experiences of people involved in traumatic and highly emotional circumstances. Such data could hardly be expected to be error-free. This criticism can be countered on four grounds.

In the first place, the sheer mass of data available (from 18 000 recorded incidents) adds considerably to the reliability of the model outcomes, and therefore to the confidence we could place in them.

Secondly, the skill of the interrogator in eliciting information from the victim for input to the model is also matched by that person's skill in interpreting the model outputs. These skills are honed and augmented by the interrogator's experience in dealing with crimes of domestic violence over many years.

Thirdly, the model outputs are not absolutes. They are forecasts of potential levels of violence if no action is taken to change the situation. The potential victim is presented with, and counselled about, alternative strategies to minimise or avoid the risk of serious assault occurring in the future.

Finally, the use of the model is not seen as a means of *preventing* domestic violence. Instead it is regarded as one important part of a number of measures aimed at *reducing* the incidence of serious assault. The model output thus joins inputs from several other sources in contributing towards decision-making for the potential victim and for the police.

The second text box (example 2.2) puts risk and uncertainty into a more every-day, personal context, and introduces some aspects of risk management. Note how any risks in example 2.2 must arise from your desire (objective) to read the daily paper and your subsequent decision to buy one during your fitness walk. Other alternatives exist. You could decide to forego the daily paper and instead substitute news and current affairs programs from the radio or television. It might be possible to access Internet news sites on your computer. You could order the paper to be delivered to your home, or buy one later in the day when carrying out other activities. You could borrow a paper from a neighbour or work colleague, or read a copy at the local library. A personal exercise machine, or a subscription to a local fitness club, could be used to substitute for the daily walk.

Example 2.2
FETCHING THE PAPER

As part of your desire to keep up with current affairs, you like to read the newspaper each day. You also want to improve your physical fitness and general well-being, and therefore decide that you will take an early morning walk to the shop to buy a paper. This daily project will help to achieve both objectives.

The nearest newsagency is 2.5 km from your house. Most of your walk can be undertaken under the overhanging awnings of commercial buildings along the way, but it does involve crossing a busy divided urban highway. The nearest pedestrian crossing, controlled by traffic lights, is at a four-way intersection 250 m further along the same road.

Alternatively, the newspaper you like is usually, but not always, available a small convenience store 3.2 km in the opposite direction. This route does not involve crossing any major roads, but there is little shelter along the way and it passes through a decaying area where the footpaths have been badly neglected.

Any of these alternatives would avoid most of the risks arising from the decision you have made, but each of them would then expose you to other, different risks. The radio or television programs might not be broadcast at convenient times. The Internet websites might not cover your favourite features. It will cost more to have the paper delivered, and delivery may be unreliable. The paper might be stolen before you can retrieve it; or it might be damaged by weather. Papers may be sold out later in the day. It might be inconvenient for someone to lend you a copy. The library copy might be incomplete or unavailable. Purchasing exercise equipment or subscribing to a health club might be financially difficult. Avoiding some risks inevitably leaves you susceptible to others.

But what are some of the risks and uncertainties associated with your decision, and how could you deal with them? We will focus on one of the more obvious ones, but think about the example carefully and you will identify a myriad of others (although you will probably overlook even more!). Do this exercise if you can, as it will help you to appreciate the need to establish the boundaries for a systematic approach to risk management. This is discussed more fully in chapter 7.

Let's assume that you decide to take the more sheltered walk, as the weather looks threatening. The first 2.5 km is straightforward as no vehicular traffic, cyclists or awkward pedestrians are encountered, and the weather remains fine and dry. As you approach the divided highway, there is still a short stretch to walk before you need to think about crossing. The traffic coming towards you, on your side of the road, is quite heavy and accelerating quickly after passing through the four-way intersection immediately before the controlled crossing. On the other side of the highway there is less traffic, and it is slowing down in anticipation of having to stop at the traffic lights.

The risk you face immediately, as a consequence of your decision to take this route, is that a vehicle collision or other incident could occur and result in one or more of the passing vehicles (or debris from them) intruding upon the walkway where you happen to be, and injuring or killing you. Given the traffic conditions and proximity, there is a slightly higher probability that this would happen with the traffic coming towards you (i.e. on the same side of the road) than from vehicles travelling in the far lanes. However, there is so much uncertainty attached to whatever the causal incident might be that it is highly unlikely that you would even think of it. For example, a passing driver might be distracted by his or her mobile telephone and lose control of the vehicle, leading to the accident suggested above. The trigger event might be so far outside the boundaries of the contextualising system (the traffic situation you are presently involved with), or be one of such a huge number of possible alternatives, and be so lacking in prior warning signals, that speculating about it would be pointless.

You are more likely to have begun to think about whether or not to cross the highway at the point opposite to the newspaper shop, or to walk the additional 500 m via the safety of the pedestrian crossing. Depending upon your abilities in mathematical reasoning, and spatial and kinetic perception, you will rapidly assess the magnitude and effects of a number of factors before making the decision. These would include the relative speeds of different types of vehicles (in at least one lane of traffic), their braking capacities and the reaction times of their drivers; traffic density (gaps between vehicles, or even a clear space caused by traffic being halted at the intersection); the need for you to walk or run (and your ability to do so, given your level of fitness!); the presence of any visual or aural impediments (something preventing you from hearing or seeing vehicles, or preventing drivers from seeing you – such as parked vehicles in the nearside lane or a sharp curve in the road, or the operation of noisy equipment nearby, or a jet aircraft passing low overhead); and perhaps many more. For each factor, in terms of both magnitude and effect, there would be considerable associated uncertainty – vehicles speeds, for example, could not be assessed with any great accuracy, and you would have to put bounds on these uncertainties in terms of what you would normally expect to encounter in such situations. All this involves further decision-making.

There is even uncertainty in your capacity to assess some factors at all, given that this capacity itself can be affected in many ways (age, mental and emotional states, etc.). So your judgment has to take in an array of situational factors, and is affected by capacitating factors. Furthermore, some factors have to be assessed and then reassessed continuously or within very short intervals of time. Remember, too, that all this decision-making is connected with the tasks emanating from your desire to achieve your fitness and newspaper reading objectives!

Your decisions are also made in the light of the alternative available to you – to walk the extra distance to the controlled pedestrian crossing. Note that the risk you have identified is the chance (probability element) that you will be struck (risk event) while you are crossing the road (duration) and killed or injured (consequence). Deciding to use the controlled pedestrian crossing will not eliminate or avoid this risk. The crossing alternative would reduce the probability of occurrence substantially, but you would still be at risk for about the same length of time, and the event and consequences would largely remain the same. In effect, therefore, your decision is about a trade-off between probabilities and the extra time taken to walk to the crossing and the same distance back from it on the other side of the road.

Assume that you have decided to take the riskier course of action, and that you have successfully managed to jay-walk (or run) to the central divide of the highway without incident. You now must deal with a similar, but not identical situation to reach the far footpath. Now you do not have the luxury of the alternative crossing for your decision, since it is no longer accessible from where you are. You have to cross the lanes of traffic proceeding towards the intersection. The magnitude of the factors will be different, bringing different uncertainties. The traffic is less heavy; it is probably decelerating rather than accelerating; there are likely to be fewer clear gaps. Any visual impediments will be different (and on-coming drivers will not be expecting to see you on their near side). You are also at a more vulnerable starting point, since in the middle of the highway you are more exposed to both streams of traffic.

The decision is made; the road crossing safely completed and the precious newspaper purchased. Now you stand at the kerbside once more, faced with yet more decisions about crossing back over the highway, but your earlier fears have been realised and a passing rainstorm starts to complicate matters. Not only will you get wet until you can reach the shelter available on the other side of the highway, but the rain adds even more uncertainty to the situation. Vehicle speeds and braking capacities will be affected; smeared windscreens or goggles will create additional visual impediments. The ambient light situation may have deteriorated. Along with all the other influencing factors, these too must enter the processes of judgment in your decision-making if your journey is eventually to finish safely at home.

Note that we have confined our exploration of the scenario presented in example 2.2 to the risk of being hurt or killed in a traffic accident. Other risks are involved in the fulfilment of the objectives sought in this scenario. The shop might not have your favourite paper. You might have been attacked by a dog, assaulted by rogues on the way, or hit by debris from a sudden building collapse, crashing aircraft, falling space junk, meteorite shower or passing parachutist. Your dangers are constrained only by our imagination!

Yet other risks would be encountered had the decision been made to take the longer, less sheltered route. All of these risks would have been subject to uncertainty associated with the factors contributing to them. Note, too, that every one of the decisions in example 2.2 would have involved communication, albeit much of it intra-personal.

Uncertainty, then, is distinguishable from risk. It is associated with risk through the inherent variability in the values of the factors to be considered in analysing each risk. True uncertainty may be converted to bounded uncertainty by assumption, based on judgment, for the purposes of risk analysis. As we shall see later, for effective risk management this must be followed by risk response and thereafter translated into the control and review mechanisms of project risk management.

Having considered risk and uncertainty from a somewhat theoretical point of view, we now turn to a typological consideration of risk.

2.4 CLASSIFYING RISK

Why should we try to classify risk? After all, if risk is a social construct, as we have argued, then each community or individual person will have distinctively different views about what actually constitutes a risk. Similarly, if the uncertainty associated with a risk is often subjectively assessed, and we know that risks must be treated individually, why bother to classify them when we could save time and effort by starting directly with the more 'hands on' activities of risk management?

Classifying risks enables us to consider them within a more coherent framework. Creating an acceptable taxonomy establishes a common basis for risk and risk management researchers to proceed with their investigations and communicate their findings. It provides us with a more uniform risk language, particularly in fields such as project management, where we might need to communicate risk information to a wide variety of project stakeholders. It allows us, therefore, to establish a common understanding of different risks, and provides an essential basis for effective knowledge transfer, within an organisation and from one project to another.

Classifying risks also provides the opportunity to explore whether a particular class or type of risk is amenable to a particular type of treatment. To date, little uniformity exists in the classification of risk, even within recognised discipline areas. Different approaches are found to classifying what are often the same types of risks.

The 1991 report of the Royal Society identifies three 'conventional' types of risk: natural hazards; technological hazards; and social hazards. Two further types are noted: health hazards and financial risk; together with the caveat that distinctions between all types of risk cannot be absolute because of cultural differences which may cause one group to perceive a hazard as natural which another group might regard as social or technological.

Under these risk categories, natural hazards are defined as those occurring outside human systems or agencies. Examples would include earthquakes and hurricanes. Hazards of technology occur within human-designed systems, and would include events such as collisions between ships or vehicles, explosions, fires attributable to electrical equipment failure, and failure of structures. Social hazards arise from human behaviours such as arson and sabotage. Health risks relate to epidemia and surgery.

Life-cycle application phases are suggested in AS/NZS 3931 (1998) as another approach to categorising risk. The phases are identified as: concept and definition phase; design development phase; construction, production, transportation, operational and maintenance phase; and the disposal phase. Little obvious advantage accrues from this as a primary categorisation approach, although for specific projects it might serve to provide a clearer focus for the time component of risk.

Yet another approach seeks to separate external and internal risks. This too is more suited to specific contexts, since a risk might be classed as external in one situation, but as internal for another. This is part of a systems boundaries issue which we discuss later in this book.

The obvious problem in classifying risk, apart from the cultural perception difficulties noted by the Royal Society report, is that there is a danger of confusing sources, causes, effects, time phases and fields of study for the risk domains. The confusion may also arise from consideration of individual risk circumstances. For example, the impact of a hurricane may be aggravated in a region where overgrazing by cattle or deforestation through deliberate land clearance has taken place (both humanly engineered situations), but essentially it is the hurricane which gives rise to the risk, so it is still a natural one in terms of its source. Similarly, while adverse soil conditions are a perennial risk for construction projects, this risk is also a natural, albeit latent, one. The potential for risk is simply activated by the decision to build on a particular site and the subsequent commencement of excavations. Yet again, the damaging effects of a soil washaway may be increased by the removal of natural water-courses during the execution of bulk earthworks excavations, but the source of the risk is still the storm which precipitated the washaway. These examples suggest that it is one or more of the elements of risk – probability, consequence and time – which may be affected by changing circumstances, not the type of risk itself.

While the source/impact dilemma remains a philosophical problem for risk classification, the source system of the adverse event may be the best primary denominator. In this case, two primary categories should suffice: natural systems and human systems, since all technological (system) and social risks must stem from some form of human activity. From this starting point, sub-categories can be developed for different risk domains. A source system approach to risk categorisation is shown diagrammatically in figure 2.3. The sub-categories of natural risks are weather systems, geological systems, biological systems, physiological systems, ecological systems and extraterrestrial systems. The sub-categories of human risks comprise social, political, cultural, health, legal, economic, financial, technical and managerial systems.

Figure 2.3 A system-based risk source classification approach

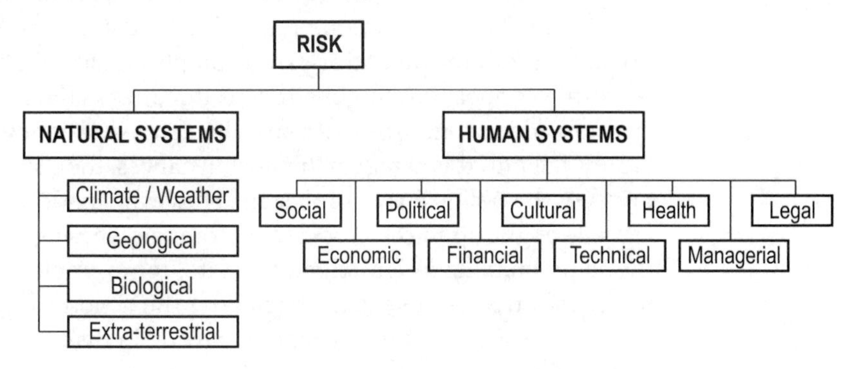

Natural risks

As noted earlier, natural risks arise from systems whose existence is beyond human agency. These may be found on or beyond the planet Earth. Weather systems, for example, are responsible for risks posed by hurricanes, typhoons, tornadoes, floods, and lightning strikes. Geological systems give rise to risks of earthquakes, tsunami, volcanic eruption, and geological faults. A meteorite shower is an example of an extraterrestrial natural risk. However, a potential collision between man-made space debris and Earth, caused by a satellite falling from its orbit, would be regarded as a human risk, in the technical category and not as an extraterrestrial system risk.

Risks arising out of the natural state of the biological diversity of species on planet Earth are also natural systems risks, but it should be noted that where effects on these systems are triggered by human action or intervention, then they are sourced as human systems risks.

Human risks

Risks arising out of human agency are more difficult to categorise than natural risks, due largely to the overlapping and interrelational characteristics of many humanly devised systems. Nevertheless, it is possible to indicate examples of human systems risks.

Behaviour which is unacceptable to society generally may be categorised as social risk. This category might include criminal acts such as theft, robbery, assault and murder, sabotage, arson and espionage; civil torts including trespass, slander and libel; or substance abuse such as drunkenness and other drug-induced behaviour. Graffiti might also be classified as a social risk, but it is important to remember that it is the values set by a society that determine which behaviours are anti-social. These values are culturally determined, and may change substantially over time for a particular society, or differ between societies.

Risks arising out of government action (or inaction) can be classified as political risks. Among these, war and abrogation of international treaties are straightforward examples. Political risks may also arise from opposition to government threat or action; as in cases of civil disorder or industrial action. The actions of lobby and protest groups fall into this category. In most instances, the risk falls upon the target of the politicians or protesters. However, innocent third parties can be caught up in the crossfire. A company leasing office space in the same building as a foreign consulate, for example, might find its business activities adversely affected by the actions of groups wanting to demonstrate against events occurring in that consulate's home nation. Similarly, the picket lines of striking workers rarely discriminate between the purposes and motives of people wishing or needing to cross them.

Cultural risks emanate from the threat of potentially negative interactions between groups of people espousing different cultural or religious values. At a government level, of course, such risks are treated politically, but this does not deny their cultural source.

Risks arising through the process of surgery or from an outbreak of epidemia may be categorised as health risks.

In the legal category of risk would be found risks arising from requirements to observe clauses relating to private contracts and agreements; as well as other legally enforceable instruments such as statutes, regulations and codes. It should be noted that, at the pre-enactment stage of public instruments, such risks would have to be regarded as politically sourced.

It is not always easy to make a clear distinction between economic and financial risk categories and, for much of project risk management, to do so may be unnecessary. One approach may be to distinguish between physical factors of production and factors affecting the cost of project finance or security of project cash flows. Thus, examples of economic risk might include labour or material supply issues; while changes in interest rates or credit ratings, or the potential unavailability of project loan finance, would be regarded as financial risks. It should be noted that the failure to achieve desired income streams, in an investment-related project, is not a financial risk but rather the consequence of other risk events.

The technical category of risk involves situations where failure of some humanly designed technical system might occur. Examples range from equipment and system breakdown or inadequacy, to collisions and accidents. While it is easy to understand the former, which are the technical risks of design failure, the inclusion of the latter two in this category may be more difficult to accept. Given that there is no deliberate human intent or presence of substance abuse (which would place them as the consequences of risk events in the social risk category), then collisions and accidents can more properly be seen as a failure (or absence) of technical systems intended to prevent them. This could also be viewed as the consequences of design failure.

Managerial risks relate to the likelihood of adverse events attributable to some failure of management. Project examples include poor productivity or excessive wastage by workers; inadequate project quality, and human resource problems such as inappropriate or inadequate staffing. Occupational health and safety risks may also be included under this category, but care may be needed to exclude risks that should more properly be sourced as social or cultural.

In the main, allocating risks to a particular category source is best done from the point of view of the organisation on whose behalf the risk management activity is being undertaken. This brings the locus for risk management more directly into focus. In the interest rate example above, treating the possible occurrence of high interest rates as a financial risk for the organisation allows a more focused approach to alternative response strategies than would the more abstract category of economic inflation. However, the niceties of fine-tuning risk categories should not be allowed to occur at the expense of the benefit-to-cost ratio of the risk management process itself!

2.5 CHAPTER SUMMARY

This chapter commenced with discussion of the concept of risk. We prefer the Royal Society (1991) definition: 'Risk is the probability that an adverse event occurs during a stated period of time.' This ensures that the four aspects of risk (probability, event, impact, and duration) are each properly considered in project risk management.

We have established that decision-making, connected with the activities, tasks, commitments and obligations undertaken, is the context for risk in project management. A distinction has been made between risk and uncertainty. A useful way of dealing with uncertainty in project risk management is to set quantitative boundaries on it in terms of acceptable limits or according to our capacity to be surprised by the eventual outcomes or values of the influencing factors.

Risks may be classified as either natural or human, according to the source system of the adverse event. Sub-categories of natural risks include events originating in weather, geological, biological, ecological, physiological or extraterrestrial systems. For project risks emanating in human systems, the source sub-categories include social, cultural, political, health, legal, economic, financial, technical and managerial systems.

Since in this book we are concerned with project risk management, the next step is to consider the nature of projects.

CHAPTER 3

LOOKING AT PROJECTS

3.1 INTRODUCTION

This chapter explores projects, their nature and complexity, and what makes them susceptible to risk. It will be argued that the nature of projects is determined by their environments and their constituent elements of tasks, technologies, resources and organisation. The environments of projects are determined by their phases of development. Each element of a project is capable of further subdivision into sub-elements, and the extent of uncertainty that these exhibit contributes towards project complexity. Complexity, however, is only one aspect to consider. Other factors which make projects 'risky' are discussed and illustrated through examples and case studies.

3.2 THE NATURE OF PROJECTS

A project is a deliberate undertaking. It is an endeavour that comprises the planned and organised achievement of predetermined objectives, usually within a given timeframe. In chapter 1 we noted that it is the pre-planned starting points and finishing points for projects that distinguish them from other undertakings, such as manufacturing or the ongoing provision of services, although the initial establishment of these, or subsequent decisions to upgrade or dispose of them, may be executed as projects.

Projects incorporate three essential elements: tasks, technologies and resources, which are brought together through a fourth element – organisation. Tasks are the project activities (what needs to be done). Technologies are the technical processes involved (how it is to be done). Resources are the means for carrying out the tasks, applying the technologies and staffing the endeavour. Organisation is the element that integrates and controls the other three. It determines who will be involved, when, and where. Decisions are made within and about each of these elements, and every decision will be susceptible to risk to a greater or lesser extent.

For projects, therefore, the context of risk and risk management lies in project decision-making. This assertion was made in chapter 2. We can illustrate this concept diagramatically in figure 3.1.

Figure 3.1 Decision contexts for project risk and risk management

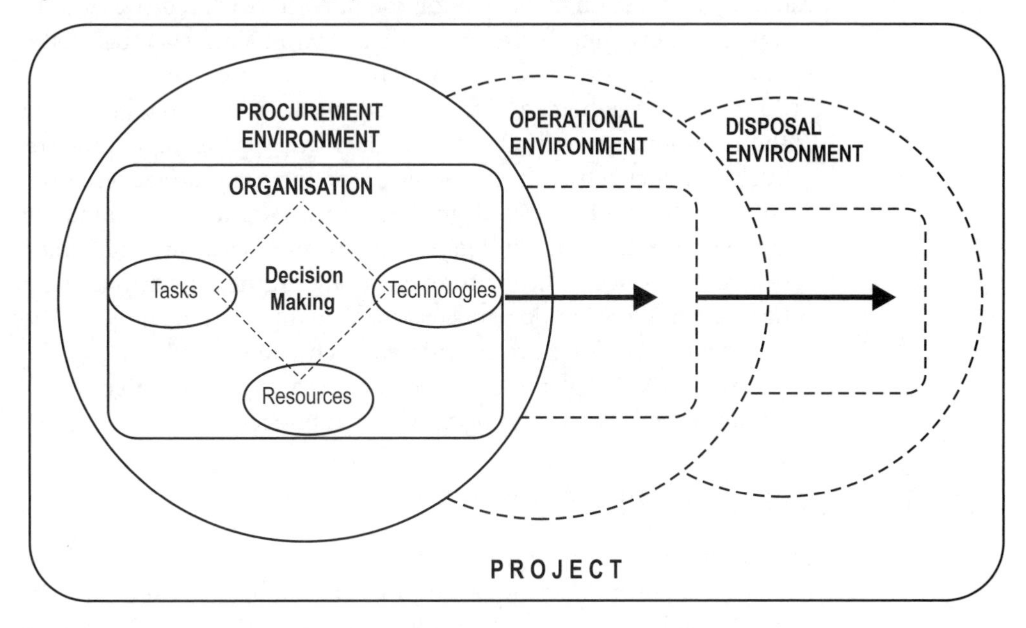

Initially, decision-making takes place within the environment of project procurement – the combination of activities, methods, resources and organisation necessary to bring a project to the point where it is ready to function as intended. Beyond the stage of the procurement environment, however, decision-making – and hence risk and risk management – may also be involved in the post-procurement environment of a project, during its operation, and beyond that to its eventual abandonment or disposal. In figure 3.1 these are depicted as 'shadow' environments to procurement, but this does not mean that they are always less important. Projects involving all three environments are not uncommon, and each may be as important as the others. In a medium to long-term situation, an open-cast mine is a good example of a project where the disposal environment (and, more significantly, its pre-planning in the procurement phase) may be as rigorous and demanding as the project's procurement and operational environments. The mining lease may have been granted on strict conditions of land remediation at the end of the period and the closure of the mine. All three environments may have been appraised, designed, planned and regulated long before the blade of an excavator first rips into the soil of the site or the ore flows along the conveyor belt. At the other extreme, events projects, such as conferences, expositions, exhibitions, parades and sporting competitions, inevitably demand consideration of all three environments on a extremely short term basis. A life-cycle approach to considering projects will expose the impacts that decisions made in the procurement environment may have upon the operational and disposal environments. Designing for energy efficiency, and for eventual disassembly are clear examples of this. The procurement environment is given prominence here, and throughout this book, simply because it is the commencing, and often the most critical, phase for all projects.

Risk is associated with all decisions made in any project environment to achieve the desired goal: in the organisation required for planning, selecting and implementing technologies and obtaining resources; or for committing to these. Many of these decisions affect other stakeholders (participants or actors) involved in the project. The effects on stakeholders are considered in chapters 4 and 5.

Given this view of the nature of projects, is it reasonable to ask whether there is such as thing as a simple project?

3.3 PROJECT COMPLEXITY

Exploring the contextual nature of projects gives us some idea about how intricate even the simplest of projects can be. Yet we frequently hear of, read about, or even work on, projects that are regarded as 'complex'. What do we actually mean by this? What makes some projects more complex than others, and what are the implications for project risk and risk management?

For this discussion, we are indebted to the work of Baccarini (1996) and Williams (1999), whose valuable contributions have extended the understanding of project complexity. Beyond this narrower project view exists the much larger field of systems complexity, which continues to attract the interest of systems theorists worldwide. That field lies beyond the intended scope of this book, but it is recommended for further study to readers who are interested in this topic, since systems theory is an important contributor to project management.

The complexity of a project is influenced by the level of uncertainty associated with two factors relating to its constituent elements and sub-elements, in any or all of its procurement, operational or disposal environments. As we have seen, any of the three environments of a project will comprise elements of tasks, technologies and resources brought together through organisation. In turn, each of these elements will usually involve a larger number of sub-elements. There will be task sub-elements, technology sub-elements, resource sub-elements and organisation sub-elements. The level of intricacy in all this is affected by conditions relating to differentation and interdependency. It is these factors that contribute to project complexity, and both are aggravated by uncertainty. Since we have already noted the association of uncertainty and risk, our interest in project complexity is understandable.

Differentiation

The condition of differentiation in a project is the degree to which its constituent elements and sub-elements should be treated as individually identifiable or, to put it more crudely, the number of distinctive parts of the project that we should recognise as such.

Note that this is not necessarily the same as the number we *could* count. In a riveted steel bridge project, for example, it might not be necessary to count every rivet. On the other hand, it may be important to distinguish between different types of rivet or between similar rivets fixed in different situations.

Remember, too, that we will not only have to differentiate between resource sub-elements (such as the supply of rivets, girders, braces and plate connectors needed for the bridge), but also between the various task, technological and organisational requirements for the project. So we may wish to distinguish between the activities of particular tradespeople; or between welding, riveting or bolting technologies; or between local and imported resources, or between the ratios of management, supervisory and line staff in stakeholder organisations. These are just a few of the many distinctions that might be necessary to make the project 'work' in the sense of being able to bring it to effective completion.

High levels of differentiation tend to complicate project planning and execution by the sheer number of different things which must be dealt with, and thus contribute to project complexity. For example, assume two similar types of projects, A and B, have the following characteristics as shown in table 3.1. Ignoring the project location factor for the moment, and purely on the basis of comparing the number of constituent elements (tasks, technologies, resources and organisations) involved; then Project B is prima facie more complex than Project A.

Table 3.1 Differentiation complexity in projects

Project constituent/element	Project A	Project B
Project location	1	5
Tasks	5	2000
Technologies	3	30
Resources	4	400
Organisations	2	15

Comparisons are not always as straightforward as this, however, since the intra-constituent characteristics of the two projects might not be sufficiently similar. Thus, although Project B requires the use of 30 technologies, and Project A needs only three, if those three are less well developed, more prone to failure, susceptible to higher variance in their outcomes, or require scarcer and more highly specialised operatives than the 30 well-understood, well-tested, well-practised and widely available technologies of Project B; then from a technological point of view Project A might be considered more complex (and certainly more risky) than Project B.

Spatial differentiation occurs in projects where elements or sub-elements have to be prepared in (or sourced from) several places, and then brought to one location for incorporation into the project. Or it may be that a sub-element is fabricated in one place, but has to be installed in many. Banking and other IT projects are a good example of this, where the centrally produced software package may have to be distributed and installed in perhaps hundreds of branch locations. It may be possible to carry out the installation loading for all branches remotely from a central place, but the project objectives will almost certainly require that the package be tested in situ so that it functions properly in each location. In terms of spatial differentiation, or indeed for temporal differentiation (i.e. project elements and sub-elements separated by time), this means that dispersion can also contribute to project complexity.

All this suggests that comparing differentiation complexity, even in similar types of projects (and the contrast between A and B in the example is exaggerated for clarity), is not easy. Comparison of differentiation complexity between different types of projects (e.g. the levels of differentiation occurring in an IT banking project compared with those in an infrastructural civil engineering project) is far more difficult, best avoided, and probably unnecessary. It can be done, but first ask why it should be done!

Interdependency

The interdependency between the differentiated parts of a project contributes significantly to its complexity and cannot be ignored. By interdependency we mean the nature and extent of any dependent relationships that may exist between the parts. If one part cannot be completed without another, or if one part cannot function without another, or one technology cannot be applied without another, or one resource is unobtainable without another, then interdependency exists. Such dependency can arise within the task, technology, resource and organisation elements and sub-elements of projects, and also between them.

It is generally held that the relationships within an organisational structure can take one of three forms of dependency: pooled, sequential, or reciprocal. The same relational argument is applicable to the task, technology, and resource elements of projects, but higher levels of interdependency complexity are found most often in the task elements. The impact of interdependency complexity is most profoundly felt in the time planning (scheduling, programming) aspects of project organisation.

POOLED INTERDEPENDENCY

If the differentiated parts required in a project element or sub-element can be dealt with one after the other, but not in any strict order, and each without interference from others, until the project is complete or a distinct point of integration is reached, then a minimum dependency relationship exists. In such a case, for example, completion times for each sub-element are simply pooled (i.e. added together) to arrive at a total time required for the whole element.

SEQUENTIAL INTERDEPENDENCY

In sequential interdependency, the conditional relationship between constituent project parts is the influencing factor. One sub-element must follow or precede another (or be undertaken at the same time) as part of an essential and deliberately planned sequence. This aspect of complexity is reflected most clearly in the critical path approach to project scheduling, where it is important to identify discrete tasks that must be completed before a following task can be commenced, as well as tasks that can (or must) be carried out in parallel. The time required for a project is represented by the length of the critical path, and not by the pooled sum of the time needed for each element and sub-element within it. In sequential dependency, however, a change in one part does not necessarily require a change in its dependent partners, as long as the sequence remains unaffected.

RECIPROCAL INTERDEPENDENCY

Reciprocal interdependency occurs when a change to, or turbulence occurring in, one element or sub-element of a project has a flow-on effect and necessitates change to, or induces turbulence in, one or more of the other elements or sub-elements. On a construction project, for example, changing the ceiling design might mean changing the type of light fittings to be fitted in the ceilings. Similarly, the discovery of a fault in a piece of program code for an IT software project might require the replacement of code in other sections.

Uncertainty and risk

How does all this complexity of differentiation and interdependency of tasks, technologies, resources and organisation relate to project risks and risk management? It lies in the level of uncertainty connected with project decision-making.

In chapter 2 it was noted that, by definition, uncertainty is some state that is short of certainty. Something is not fully known; information is incomplete. In projects, many decisions are made on the basis of forecasts of outcomes or events occurring at some time in the future. The future cannot be known with certainty. Therefore each forecast is vulnerable to some degree of uncertainty in terms of the input factors to the model used to produce it (or to uncertainty in the performance of the model itself). Depending upon the nature of the forecast, this uncertainty may relate to the likelihood of each event occurring, its timing, or the magnitude of any consequences. Since these are the essential components of risk, as we noted in chapter 2, then project risks arise out of the uncertainty associated with the decision-making relating to the task, technology, resource and organisational requirements of projects.

As we know, most projects are subject to at least some degree of uncertainty, but uncertainty states are rarely static. The level of uncertainty in a project changes, both in nature and degree, as time passes and project elements and sub-elements are completed or changed to suit new circumstances. The relationship between project complexity and uncertainty is therefore dynamic. There is an obvious paradox in all of this. One approach to dealing with large complex projects is to split them up into smaller, more manageable parts. Doing this, of course, not only increases the overall level of differentiation complexity, but it may also create new or different interdependencies and add to uncertainty.

Figure 3.2 illustrates the multidimensional and dynamic nature of complexity in projects, if at the cost of some oversimplification. Only the organisation element lists potential sub-elements, in this case different areas of management likely to be found within the project stakeholder organisation. None of the other elements displays any the subdivision that would certainly exist in real life. The diagram is useful for understanding the concepts of complexity and uncertainty, and how they are associated, but it is less effective in portraying their dynamics, since these will inevitably change over time. Nor does it tell us specifically about complexity and risk in terms of real-life projects. For that purpose, risk mapping may be a more practical approach.

Figure 3.2 Project environments, elements and complexity factors

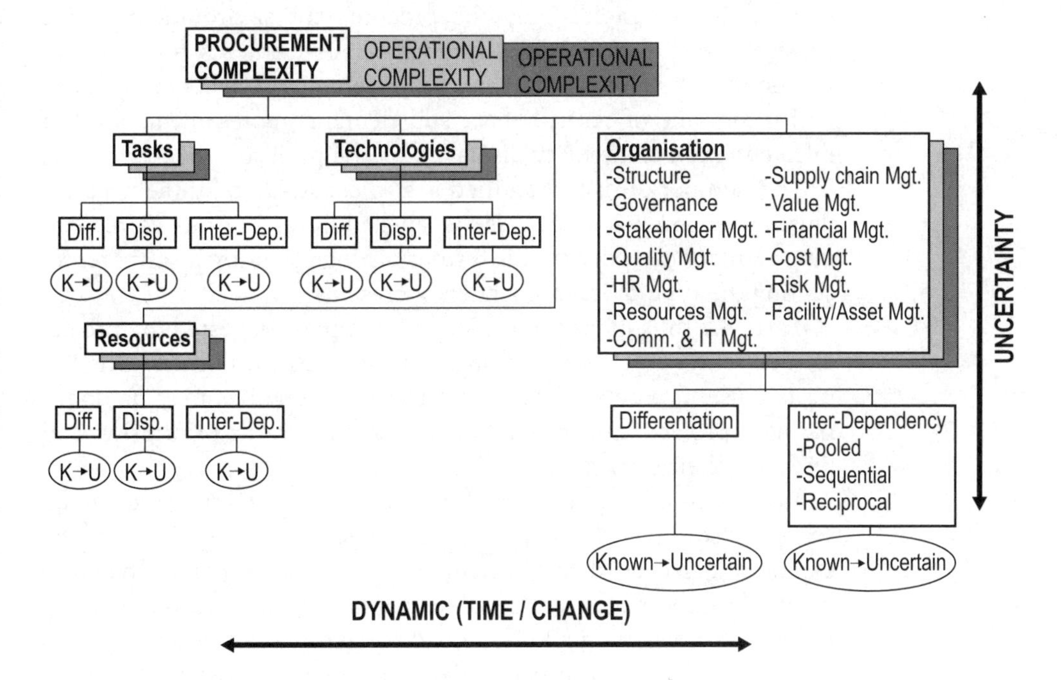

3.4 MAPPING PROJECT RISKS

The difficulty of comparing complexity in projects has already been mentioned. In many instances, it may be neither practical nor desirable to make such comparisons on an absolute, quantitative basis. The relative complexity of projects then becomes a matter of subjective assessment, and susceptible to all the errors and biases of human judgment. Given these limitations, however, it is possible to devise techniques to compare risks on an inter-project basis.

Assuming that we have, as part of an organisational risk management system, completed a process of identifying the risks for a particular project (we will look at this in chapter 6), then it should be possible to map these risks against the specific elements and sub-elements of the project, and against agreed risk classifications such as we discussed in chapter 2.

The benefit of mapping risks in this way is that it is then possible to see more clearly how risks (and the way in which particular types of risks) are distributed throughout the project. A valuable picture of the 'riskiness' of the project is obtained, which can be compared with the risk maps for alternative projects or completed projects. To some extent, this overcomes the impracticality of making quantitative comparisons of complexity.

A hypothetical project risk map is set out in table 3.2. This, too, is a simplified example. Only the procurement environment of the project is included, and the number of sub-elements for the task, technology, resource and organisation elements is deliberately limited for the sake of clarity. These sub-elements represent the level of differentiation complexity displayed by the project.

The number of risks identified is also constrained by the need for illustrative simplicity. Each risk is labelled in terms of its corresponding element category, and is further distinguished by a suffix representing the type of interdependency present.

Table 3.2 thus maps the risks in the manner suggested: by project element and sub-element and against the categories of risk. The latter is a useful way of showing which risk categories may be dominating a project, and thus act as a pointer towards prioritising and resolving their treatment.

Taking this concept one step further, if the labelled risks were to be scored according to their individual severity (this notion will be explored in greater detail in chapter 7), then some idea of the comparative riskiness of projects – or the comparative riskiness of parts of individual projects – could be gained by summing these scores.

At this point, it is useful to reflect on the characteristics of projects that may cause them to be considered as risky. In other words, what particular aspects of projects might cause them to display significant clustering on their risk maps?

3.5 RISKY PROJECTS

Apart from some measures of financial risk, the 'riskiness' of projects – like the complexity of projects – is relative rather than absolute. We cannot easily say that Project A is 6 units risky on a scale of 10, and that project B scores 4, therefore A is 50 per cent more risky than B; we can only infer that Project A appears to be a lot more risky than Project B. Our judgment will be based upon factors that we may know can contribute towards risk in that particular type of project. Our knowledge may come from experience, training, reading, or from the analytic information that project risk maps can provide. Even where risk severity has been assessed in terms of the scoring approach suggested above, the results are likely to be too imprecise to permit absolute distinctions to be drawn.

For financial risk, mathematical measures such as yield (internal rate of return), net present value and payback period can be modelled and compared across similar types of income-producing projects or even against completely different types of alternative investment opportunities.

Table 3.2 Mapping project risks into project elements (procurement environment)

Differentiation		NATURAL SYSTEMS RISKS				HUMAN SYSTEMS RISKS								
		Weather	Geotechnical	Biological	Ecological	Cultural	Economic	Financial	Health	Legal	Managerial	Political	Social	Technical
Procurement														
TASKS	A1	AR1p						AR2p						
	A2	AR1p						AR2p						AR3s
	A3							AR2p						AR3s
	A4							AR2p						AR3s
	A5										AR4r			AR3s
	A6										AR5r			AR3s
	A7		AR6r					AR2p						
	An		AR6r					AR2p						
TECHNOLOGIES	B1	BR7p						BR8p						BR9r
	B2		BR10s					BR8p						BR11s
	B3							BR8p						BR12s
	B4							BR8p			BR14s			BR13s
	B5													
	Bn													
RESOURCES	C1						CR1r	CR2p			CR4s			
	C2						CR1r	CR2p			CR4s			CR11p
	C3	CR10p			CR9p		CR1r	CR2p			CR4s	CR5p		CR11p
	C4			CR8p	CR9p		CR1r	CR3p	CR7p		CR4s	CR5p	CR6p	CR11p
	C5			CR8p				CR3p	CR7p			CR5p	CR6p	
	Cn													
ORGANISATION	D1					DR6p				DR4r	DR1p		DR2p	DR3s
	D2	DR7s				DR6p			DR5s		DR1p		DR2p	
	D3	DR7s				DR6p			DR5s	DR4r	DR1p			
	D4					DR6p			DR5s	DR4r				
	D5									DR4r				
	Dn									DR4r				

Key: p = pooled interdependency; s = sequential interdependency; r = reciprocal interdependency

Other types of risks are less susceptible to such straightforward mathematical assessment. Even the outcomes of financial risk calculations are rarely known with any reliable degree of certainty, since they are the outputs of models which depend upon predicted values for much of the performance input data. Some data may have to be transformed into financial terms. For a commercial property investment project such as a suburban shopping centre, for example, variability may exist in any or all of model input factor data including: land price (if not already owned); construction cost; design efficiency; rentals obtainable (over the whole investment period); vacancy rates; marketing costs; legal costs; insurance premiums; frequency and costs of maintenance, repairs and replacements; timing and cost of refurbishment; local authority rates and taxes; and many more. Thus, there will always be limits to the precision by which the financial risks of alternative investment projects can be analysed and compared. In chapter 2 we also noted that there may be limitations in the accuracy and reliability of the models themselves.

What then, are the conditions which are likely to make a project be regarded as risky? Cooper and Chapman (1987) and Smith (1999) have proposed factors relating to the riskiness of construction projects, but most of these factors are applicable to projects generally. According to these authors, a risky project might be one in which any of a number of conditions exist. The following list is summarised from the work of these authors, and is not presented in any particular order of importance:

1. Large capital outlays are involved.
2. Unbalanced cash flows are likely to occur.
3. There are substantial requirements for new technology (or technology which is new to the user).
4. Novel or unusual procurement arrangements are contemplated.
5. Novel operational requirements are intended by the client.
6. The project is extremely large.
7. The project is highly complex.
8. Severe time constraints exist.
9. Some or all of the stakeholders are inexperienced.
10. The client's business is highly sensitive to the performance and/or quality of the project.
11. Stringent, inconsistent, or changing regulatory requirements are encountered.
12. Environmental or ecological sensitivity is encountered in the procurement, operational or disposal environments of the project.
13. Political and/or cultural sensitivities are significant.
14. Situational turbulence is encountered (e.g. projects in developing or politically unstable countries).

Not mentioned in this list of factors, but just as important, would be the effectiveness of the stakeholders' risk management. If the risk management is ineffective, or non-existent, then the potentially adverse impacts of many risks will be exacerbated. Because little or no risk mitigation is undertaken, the likelihood of occurrence for some risks, their consequence, or the period of exposure, may also increase.

As noted earlier, few of these characteristics are susceptible to precise measurement, and their effect on project riskiness is therefore seldom known with any accuracy.

Large capital outlays

Large capital outlays generally affect the risk level of projects through the need to acquire funds beyond a stakeholder's ability to finance a project from equity. Very large engineering projects, such as dams, power stations, tunnels, major freeways, etc., may have budgets amounting to hundreds of millions – or even billions – of dollars. Their financing may create turbulence in capital markets, and loan security may be of a non-recourse nature, i.e. based solely on forecast future income streams from toll revenues or output purchasing agreements. The sheer financial size of these projects means that they may lie beyond the 'swings and roundabouts' dampening effect that most financial markets rely on for stability over time. Their size also means that the financial markets for such projects might be quite restricted, and a particular market might refuse a loan for a particular project simply because the lenders are already engorged with loans on other, more attractive projects. This would force the borrowing project developer to seek finance in alternative, unfamiliar markets, thus raising the level of risk.

Unbalanced cash flows

Long-term projects which include an income-producing operational environment tend to rely on the regularity of the operational positive revenue streams to counter the large irregular negative cashflows normally required in the procurement phase to bring the project to operational readiness. If the operational revenue or expense streams are likely to be very irregular, or the project requires a massive cash outflows in its final disposal environment (such as the open-cast mine example mentioned earlier in this chapter), then it may be extremely difficult to model its financial performance with reliable accuracy.

Highly unbalanced cash flows can actually affect the performance of computerised discounted cash-flow models used in financial appraisals to calculate an internal rate of return or yield for a project. Further, the appearance of project instability which such cash flows portray contributes to a perception of project riskiness which may not be fully deserved. In many markets it is the perception of performance that counts.

New technology requirements

New technology always carries a higher risk of failure than technology that has been tried and tested. In a major transport infrastructure project in Australia, for example, the innovative electronic tolling operational system for road users gave rise to protracted problems in both its development and commissioning stages, and lead to delays in opening the road. Government pressure, stimulated by public protest, forced the BOOT (build-own-operate-transfer) joint venture concessionaire organisation to open the road on a toll-free basis for several months, thus affecting the forecast revenue streams from the project.

It is not even necessary for the technology to be new per se. Existing technology, but which is new to the project user, can have a similar risk effect. In this case however, there should be less risk of technical failure, as any faults or weaknesses should have been eradicated by previous users.

Novel procurement methods

In regards to novel procurement methods, procurement is understood as the contractual arrangements defining the rights and obligations of the parties directly involved in a project, and the relationships between them. These arrangements may take alternative forms, but the differences usually relate to the ways in which specific risks are allocated between the parties. This is a more limited understanding of procurement than the definition we assigned earlier as the total set of activities necessary to bring a project to the state of operational readiness required by its owner.

With this in mind, it should be noted that there are few completely 'new' methods of procuring projects. Instead, a method already used in one particular field may be taken up in another, as its benefits are perceived to be greater than the method currently in use in that field. Or a system practised in one country will find its way to another through international trade activity, or perhaps even through the migration of people with the requisite experience.

The novelty in alternative procurement methods lies mainly in the unfamiliarity experienced by the people engaged in administering them. In the UK construction industry, for example, one building procurement system in use since the early nineteenth century has come to be regarded as traditional, simply because it was widely adopted in both public and private sector projects over nearly two hundred years, and is still popular today. Under this system, design and construction are clearly separated. An owner engages an architect to design the building, often with the assistance of other professional consultants, and to prepare tender documentation. A period is then set aside for tendering. Bids from contractors are submitted to a process of adjudication, and the owner enters into a formal contract agreement with the successful tenderer, who then proceeds to construct the building according to the architect's design. Usually, the architect is retained by the owner to supervise the administration of the contract and ensure the sufficiency of the work carried out by the contractor.

Minor variations in this UK system have developed over the years, but the basic principle of separating the design process from the construction process has remained intact. In the wake of British colonialism, the separated design-tender-construct system has also flourished in most of the countries of the former British Empire.

In the mid-twentieth century, financial pressure on client bodies to contain building costs saw the introduction of an alternative building procurement system under which design and construction became integrated. Using this approach, clients called upon contractors to submit tenders for the design and construction of building projects. The 'design/build' method, as it became known, removed one problem associated with the traditional approach. The increasing complexity of buildings, particularly in terms of their services installations, had begun to put the design of these components beyond the single capability of many architects. More and more frequently, contractors were called upon instead to appoint specialist subcontractors to design and install the required services systems. Although attempts were made to remedy the breach with special clauses in contract agreements, this development began to bridge the design responsibility gap that had been carefully put in place under the traditional separated procurement system. The 'new' design/build method placed all design responsibility (and hence the technical risk of design failure) clearly in the hands of the contractor, even though in practice a contractor might then employ external consultants to undertake the design work.

Curiously though, the design/build procurement system was not new. It was already well known and commonly used throughout the rest of Europe, and universally used across the United States of America. In fact, nor was design/build really new in Great Britain. Prior to the adoption of the 'contracts-in-gross' system of the eighteenth century, which was developed to cope with the rapid urban and industrial expansion then taking place, and which itself was the precursor and inspiration for the development of the separated procurement system, the great majority of construction work in Britain had been procured in a manner very similar to design/build. Rich clients would commission architects to design and build their houses, with the architect taking responsibility not only for the design, but also for engaging and paying tradesmen and for purchasing the requisite materials. In effect therefore, modern design/build procurement systems are simply closing the circle, at least in the United Kingdom.

The same applies to other 'new' forms of procurement. Much is made of the novelty of the 'Private Finance Initiative' (PFI) for delivering public services and goods in the United Kingdom, but similar forms of partnership between the public sector and the private sector, where the public client divests itself of the financial responsibility for such delivery by transferring it to private sector organisations in return for regular payments or a share of the revenue streams, have been used in other parts of the world for upwards of fifty years. The main difference is the scale and frequency with which these public/private partnership procurement systems are now employed, and the range of projects where they are found.

So it is more likely to be the novelty of the procurement method for the *stakeholders* that gives rise to risks in a project, particularly where this involves untried financing arrangements.

Novel operational systems

Technically novel operational systems for projects are more likely to be encountered than administratively novel operational systems, simply because the rate of technological change generally exceeds the rate of organisational change (and organisation is what administrative operational systems rely upon).

Examples of novelty in operational systems can be easily found. The technical novelty of the operational tolling system for an Australian BOOT type road project has already been mentioned. Organisational system novelty was evident in PFI concessional arrangements for some school projects in the United Kingdom in the late 1990s and early 2000s. In a BOT (build-operate-transfer) type arrangement, private sector sponsors agreed to build and maintain school facilities in return for regular fee payments over the concession period from the public sector clients. As with novel procurement systems, it is the novelty of the operational system to the user which gives rise to greater risk. Public sector expertise for administering and maintaining school facilities was built up over decades in the United Kingdom. That expertise was not immediately available to the private sector operators under the new PFI arrangements. These operators, mostly originating from the larger construction contractor organisations, had to develop their own facilities management systems from scratch. To a large extent, such organisations were unfamiliar with this aspect of projects, particularly with the extended time scales involved compared with the relatively shorter periods required for just constructing the facilities. It is hardly surprising that several of these private sector organisations are now looking with a very critical eye at the long-term operational obligations to which they have committed themselves.

Very large projects

Very large projects can be risky because of their sheer scale. For example, they tend to be highly resource dependent and thus vulnerable to turbulence in the supply of these resources. A very large project can exhaust the capacity of local markets to supply its needs, creating logistical bottlenecks and driving up prices. Labour unions may hold the project to ransom, knowing that alternative sources for labour are scarce or non-existent.

The length of time required for development can also act against very large projects, as momentum may be lost that is difficult or impossible to recover, leading to delay or even abandonment.

Highly complex projects

The complexity of projects, and how this contributes to risk, has already been discussed. It is important to stress that complexity is not necessarily synonymous with size. Small projects, with high levels of differentiation and sequential and reciprocal interdependency, may be far more complex than large-scale projects containing few elements and straightforward pooled interdependencies.

Severe time constraints

Most projects have tight time-lines, and it seems to be part of the intrinsic nature of all projects that they are difficult to complete within a predetermined time. Delays and extensions of time are the norm rather than the exception. Whether this is due to inadequate preplanning of project schedules and programs (a management risk); or to deficiencies in project supervision and control (also a management risk); or to other causes, is a matter of individual investigation for each project. In some cases, while the impacts of completion delay may be inconvenient, they are often not critical.

A delay of a week in opening a new shopping centre, with the event scheduled at the end of the month to exploit the opportunity to attract monthpaid shoppers, might be disappointing rather than disastrous, and would almost certainly not be critical in terms of the longer term financial profitability of the project. If the delay were to occur just before pre-Christmas opening planned to exploit the bumper sales season, however, the impact on sales cash flow might be far more serious, to the extent of critically endangering the ability of smaller retail tenants to survive the high start-up costs of their participation. The loss of the 'kick-start' event might then affect the longer term attractiveness and status of the whole centre.

However, examples such as these are at the lower end of the risk continuum for 'time to market' projects, in terms of their need to be completed in time to capture market opportunities. Other types of projects have far more urgent time constraints. Nowhere is this seen more clearly than in the telecommunications industry. Projects in this field, especially in mobile cellular communication, can be as short as four to five weeks in terms of preparation and execution, in contrast to the three or more years usually needed for a new suburban shopping centre. ICT (information and communication technology) projects such as introducing new call charge plans or offering additional telephone services are usually aimed at increasing or maintaining market share by beating competitors to the punch. Time is therefore a critical factor for them.

For other projects, failure to achieve completion by due date may be extremely serious, or even catastrophic. With such projects, the predetermined completion date is the most critical criterion in terms of project success or failure.

In Australia, the introduction of GST (general sales tax) legislation required commercial and other enterprises to be ready for its implementation on the 1 July 2000. Business enterprises were faced with the need to change pricing structures, adapt cash registers and accounting systems, and train staff. Some organisations, such as charities, found themselves exposed to such requirements for the first time, since they had not been subject to the preceding, more selective, sales tax regimes. All this created a myriad of change projects, and a huge last-minute demand on the resources needed to execute them. As a result, many small organisations made unwise purchases in terms of software and systems, simply to meet the GST implementation date, and were still trying to rectify the inadequacies of these two years later.

Such 'fixed end' or 'fixed time' projects are internationally and supremely exemplified in the worldwide apprehension prevalent in the 'Y2K' period just before the end of the second millennium. Few will forget the horror stories of what might happen at midnight on 31 December 1999. Because of a potential inability of older computer systems to cope with a double zero digit year in the formatting of calendar dates (e.g. 01/01/00), information systems failure on a massive scale was predicted. Aircraft might fall from the sky, trains might run on the same tracks; emergency services might be rendered useless; stock markets might have to cease trading – the list was endless. On the other hand, plans to counter the threat also received great publicity. Some airlines planned to keep their aircraft on the ground during the critical period. One nation's army command decided to wind back the dating systems of all its computers by a decade. Another organisation, perhaps more subtly, decided upon the same ploy but went back only the number of years needed to ensure that the days and months matched the year 2000 calendar.

The great majority of Y2K projects, in the three or four years leading up to 2000, involved ascertaining the compliance level of computerised information systems – testing them well before the critical time to ensure system integrity and capacity to cope with the date format issue. Systems and equipment manufacturers and suppliers also did unprecedented business in replacing suspect installations – thus introducing unforeseen life-cycle risks to the original installation projects. All this placed enormous pressure on IT resources worldwide.

The demand for compliance testing had three potential consequences. First, existing IT staff resources might not be sufficient to complete project tasks in time, thus increasing the risk of Y2K system failure. Second, less competent entrepreneurs might try to exploit the market opportunity and then fail to carry out the compliance testing tasks properly, again increasing the chance of system failure. Third, and worst of all perhaps, clients might be deterred by the increasing costs of IT compliance projects (attributable largely to the human resource scarcity) and decide to minimise, or even ignore, any form of system compliance testing. Trusting to luck in this way was probably the worst form of risk management to adopt.

As yet, there is no clear evidence that any serious system failure, which could be attributable to one of these consequences, actually occurred in the early hours of 1 January 2000. The benign reality of Y2K, in contrast to the catastrophic nature of its earlier anticipated threat, has ensured that any inadequacies in Y2K compliance projects have remained well hidden. The lesson from the Y2K experience is very clear, however. For 'fixed time' projects, adequate prior planning and scheduling are critically essential for project success and for mitigating risk.

Stakeholder inexperience

Project failure due to stakeholder inexperience is found, more often than not, in the technology and organisation elements of projects. Most project stakeholders know what is to be done (the tasks), and what will be needed (the resources), but how the tasks should be undertaken (the technologies) and who should be involved and when (the organisation) can often prove far more difficult to envisage and arrange.

Performance/quality standards

The higher the standard of performance or quality required in a project, the greater is the chance that completely satisfactory outcomes will not be achieved.

If digital imaging technology using overhead gantries, for example, is to be used for capturing the licence plate details of passing vehicles as part of a toll-road charging system, then the resolution of the image must be sufficient to allow human interpretation of numbers which do not match those held on a database for vehicles whose drivers have prepaid the toll. Similarly, for an electronic road pricing (ERP) project in Singapore, aimed at reducing traffic congestion by discouraging private road users from entering central city streets, it was found that the heat and strong sunlight affected the integrity of the smart cards in the tags located on vehicle windscreens. Higher quality plastics had to be used to resolve the problem.

Regulatory environment

Projects carried out in a highly regulated environment run the risk that, at some point, something or someone connected with the project will be in breach of a law or regulation. However minor the breach is, it may have the capacity to cause a major disruption to the project. There is also the chance that laws and regulations might be changed during the life of the project and similarly cause disruption.

Since no project takes place in a completely regulation-free environment, it might be said that all projects are risky in this regard, and their vulnerability is simply a matter of degree. Even so-called 'enterprise managed projects', which are projects carried out wholly within an organisation, will still be subject at least to the internal rules and policies of the organisation, and are likely to have to comply with numerous external regulations relating to matters such as occupational health and safety, employee award determinations, permits and licences.

Environmental/ecological sensitivity

Sensitive environmental conditions and issues can complicate an otherwise straightforward project. As with other factors, the risks may arise not only in the procurement stage, but also in the operational or disposal phases of a project. Nor are the risks confined to the sensitivity of purely physical environments. Since the adoption of 'triple bottom line' objectives for many projects nowadays, the implications of their social and economic environments may also have to be considered in terms of risk. The impact of a project upon its surrounding community, or upon local, regional or even national economies, may become critically important to the point where these can affect the feasibility of the whole project.

Ecological sensitivity is a growing area of concern, matching the parallel growth in ecological research and discovery. On some projects, precautions against ecological damage have had to be changed or even abandoned after new findings have shown them to be ineffective. Environmental and ecological issues also have the propensity to arouse activist and protest lobby interventions, thus giving rise to additional political risks.

Political/cultural sensitivity

Few projects take place in environments that are entirely free from political or cultural sensitivities. The level of these sensitivities may range from intra-project issues and differences, through those at intra- or inter-organisational levels, to even higher local, regional, national and international sensitivities.

Projects which, by their very nature, intrude upon or exacerbate political or cultural sensitivities are clearly most at risk. However, it is possible for a project to be innocently caught up in such situations. It is not unknown, for example, for prominent or urgent construction projects to be targeted for industrial action by unions eager to demonstrate their power to governments or employer groups unwilling to accede to their demands. Delivery of internationally sourced project resources may be held up at ports or airports for similar reasons, or for the purpose of showing solidarity with unions overseas. Nor is it unusual for governments to delay projects until after elections have taken place, in order to avoid losing key votes. The 'pork barrelling' opposite, where projects are accelerated to appease influential groups, is also well known.

Multicultural development in many westernised societies has brought with it a parallel increase in the need to test the cultural sensitivities of many projects in terms of their potential effects on accepted religious and ethnic practices.

Situational instability

Situational instability is most often interpreted as the additional risks of undertaking projects in foreign countries, particularly in third world nations. The instability usually derives from a number of the factors already discussed, particularly environmental, political and cultural sensitivities, novel procurement systems, capital financing and stakeholder inexperience.

Other situational factors might include the sanctity of government guarantees, the ability to remit profits, and the safety and welfare of expatriate employees.

Project examples

The examples shown in text boxes show how some of the factors that we have discussed here can affect the level of risk in projects.

SHIP MONITORING PROJECT

In example 3.1, the island radar tower installation project for monitoring shipping is obviously vulnerable to greater levels of political, legal, cultural, environmental and technical risks than a comparable project constructed at an urban mainland location.

Example 3.1
SHIP MONITORING PROJECT

The Austrialian Maritime Safety Authority (AMSA) was concerned about the increasing number of ships using the environmentally sensitive inner passage of the Great Barrier Reef on Australia's eastern seaboard. Together with the Queensland Department of Transport, the AMSA initiated a project to install a radar and automatic identification system to track shipping. The site selected for the installation was on a remote island. The nearest mainland town of any size is Townsville. Although uninhabited, the island presented legal and cultural (in terms of original and traditional ownership rights) as well as environmental sensitivities.

Project stakeholders included federal and state government departments, traditional landowners, environmental protection groups, contractors, and subcontractors specialising in construction, transport and communications. Extensive consultation and negotiation was necessary, before and during the project, thus increasing the need for effective stakeholder management and communication.

Barge and helicopter transfer was required for landing materials. Access to the site from the beach was limited, and hand clearance was necessary as it was not practicable to mobilise heavy machinery. The 'footprint' of the permanent 15 m high tower structure, and the working area for the site, had to be kept as small as possible to minimise disturbance to flora and fauna. Disposal of construction, packaging and other waste, and the final clean-up after completion, necessitated careful planning and execution. Weather delays were a major consideration. In the technical design of the installation, the need to prefabricate components wherever possible, together with the desire to minimise ongoing maintenance requirements (thus reducing the frequency of subsequent visits for systems servicing), were important constraints.

From inception to completion, the project took approximately 30 months.

Careful negotiation and consultation, in the preliminary phase of the ship monitoring project, reduced the probability of occurrence for the political, legal, cultural and environmental risks associated with it. More detailed design and operational planning (than is normally the case for such projects) was undertaken to reduce the probability of occurrence, and the potential impact, of the technical risks of this combined construction and IT-related project. The 'cost' for these reductions was the additional project procurement time needed – perhaps as much as 6 to 9 months longer than a similar project constructed under less difficult conditions. Another measure adopted to reduce the probability of occurrence of technical risks was the trial assembly of critical components (e.g. the tower structure) on the mainland prior to transporting them to the island site. In this case, the risk management cost was the direct cost of the trial assembly and disassembly, and also the additional time required.

The client organisation decided to station a project inspector almost permanently on-site during construction, in order to facilitate ongoing design decision-making. On the one hand, the technical risk event would have been construction failure (albeit not serious), with its consequent costs of delay and re-work. On the other hand, the availability of the project inspector reduced the likelihood of occurrence of the technical risk of design failure, but the risk management cost was the additional cost of this supervision.

This example shows the nature of what might be considered a risky project, particularly in terms of risk factors (3) new technologies, (10) stringent quality or performance standards, (12) environmental and ecological sensitivity and (13) political and cultural sensitivity, in the list discussed earlier. It also points to how this riskiness may be addressed through appropriate risk management.

COLLEGE IT CENTRE PROJECT

Example 3.2 demonstrates how some risks are more subtly shaped. This example also refers to a combined construction and IT-related project. However, less evidence of careful risk management is found in this example.

Example 3.2
COLLEGE IT CENTRE PROJECT

A college in South Africa was concerned about the level of computer literacy of its students. The educational neglect of the previous apartheid government regime, together with regional poverty, had left students poorly prepared both for the demands of tertiary study and for the keenly competitive job market.

The college council decided to build a new IT centre on a vacant campus site. The project objective was to create the physical environment and facilities for raising the level of computer literacy among students.

A traditional building procurement system was adopted for the project, using a separated design–competitive tender–construction contract approach. The college council had used this procurement system on all campus development projects to date, and was unfamiliar with alternative approaches.

Piling for the foundations of the new building would be carried out under a preliminary direct contract with a specialist contractor, followed by a separate competitively awarded contract for the main construction work. A further separate negotiated contract would be entered into with a specialist company for the design and installation of a tensile membrane roof over the atrium of the new building.

The project design team comprised professionals appointed from preferred lists of architectural, engineering and quantity surveying consultants qualified to undertake public sector work. Project management was undertaken by the college itself, using staff from its campus planning division.

One variation to the traditional procurement system was incorporated for the project. Under the affirmative action policies of the post-apartheid South African government, professional consultants engaged in public sector commissions for building design work are required either to demonstrate a racial balance in their senior management and staff employment, or to be willing to work under a joint venture type of appointment, whereby they would provide the design services in tandem with smaller firms wholly comprised of staff from previously disadvantaged population groups. For the college IT centre project, the latter arrangement was used for the architectural and for the mechanical and electrical engineering consultancies. The structural engineering consultant and the quantity surveying consultant already had the requisite affirmative empowerment measures in place in their organisations, as did the main contractor. Had the main contractor been unable to comply with the policy requirement, its tender would have been penalised against all complying competing bids.

The traditional procurement approach was chosen because it allowed the project design to be developed in step with an evolving project brief. A complete and detailed brief had not been possible at the start of the project (compared to other campus buildings whose spatial and functional requirements could be identified and quantified with some precision) since no comparable facilities existed on the campus or elsewhere. The new building was designed simply as a three-storey shell totaling 6500 m^2 of floor space, with an external appearance to match its neighbours. When the resource requirements for the IT elements of the project could be known with greater certainty (through extensive consultation with college staff and assessment of students' needs), more detailed internal spatial planning and equipment design would be undertaken. The initial brief, based upon rudimentary surveys, called for a building capable of accommodating 1500 work stations for unstructured – as well as tutored – student use; plus a literacy and writing centre, a video-conferencing unit and an audio-visual suite. It was accepted that these parameters might subsequently change.

A capital cost of ZAR 26 million was estimated for the building project, including allowances for escalation, sales tax charges and professional fees but excluding IT-related equipment, furniture and software. Unlike previous campus development projects however, the government education authority would not provide 100 per cent funding for the IT centre project and a substantial shortfall was envisaged. Nevertheless, the college council decided to proceed, confident in its ability to raise the necessary additional funds from alternative sources.

Construction work was scheduled to take 14 months.

Most obviously, perhaps, the lack of a precise brief for the college project incurs the risk of functional inadequacy of the completed facility. It could be argued, though, that the construction of a 'shell' is a risk mitigation measure, since this would produce a facility flexible enough to accommodate a reasonable amount of spatial changes and changes in information technology. Yet developments occurring in IT show how this view might be too narrowly focused. For example, a similar educational organisation in Singapore has installed 'wireless' computer interface systems to serve its campus. Students can hire laptop computers from the college (or use their own), connect to the Internet from almost any location in the college buildings, and access a huge range of IT resources including tutorial help with computer applications. The 'wireless' innovation actually supercedes an earlier system of fixed wall outlet points installed in selected areas of the campus (student lounges, corridor rest areas), as these were found to be unable to cope with the student demand and expensive to retrofit.

The South African college, however, has locked itself into an expensive short-term solution to a problem which in itself is dynamic, since the computer literacy of incoming college students is largely a function of their prior exposure to computers. Even allowing for the low base engendered by the socio-economically disadvantaged backgrounds of many South African students, it is likely that the computer ability of incoming students will rise rapidly, and over months rather than years. Such is the nature of IT and its users.

One small area of risk on this project relates to the separate contract for the piling. In order to avoid excessive noise nuisance, this work has to be commenced and completed within the period of the long summer (southern hemisphere) vacation for the college. Should unforeseen ground conditions or other factors cause a delay in completing the piling, the impact might lead to an intolerable noise situation for students and staff; costly weekend overtime work; or at worst a cessation of work until the following Easter vacation. This would then have a knock-on effect in terms of completing the whole project in time for start of the following academic year.

A less obvious indicator of the comparative riskiness of the college IT centre project lies in the financial risk of exceeding the budget. The lack of precision in the project brief suggests that the likelihood of this happening is high.

Generally, if a construction project budget is very tight, a client should be advised to employ a procurement system that will guarantee a fixed price – a design/build or package deal approach, for example, where the contractor takes full responsibility for design and construction, and a price is determined on that basis before construction is commenced. This approach, however, requires the client to provide a very precise brief of his or her requirements. That avenue was clearly not possible for the college. Even if a precise brief had been available, the college was not familiar with any procurement system other than the traditional separated design – competitive tender – construction approach, so that adopting an alternative system would have created its own risks of inexperience.

Subtle risks in this example also lie in the adoption of a tensile membrane roof design for the building's atrium. The decision was based upon aesthetic and symbolic grounds, rather than for practically functional reasons. This type of roof was thought to symbolise the cutting-edge nature of the building's purpose (despite the historical longevity of membrane roof technology!). In the first place, neither the client nor their consultants were familiar with this particular roofing technology, hence the need to rely upon the expertise of a single specialist company to design and install it. Any failure on the part of this company would leave the college vulnerable to expensive delays and re-design costs. More importantly, perhaps, this design feature had already been singled out privately by the construction cost consultant as being dispensable if project cost overrun became likely later on. While this appears a sensible precaution, friction between the consultants could arise if the possibility of omitting the membrane roof became known to the project designers only late in the construction phase. At best the consequence might be a communication breakdown between the project consultants at a critical time for the project financially. At worst, open conflict could ensue.

The joint venture arrangements for the project consultants also pose risks. The affirmative empowerment procurement policy in South Africa is intended primarily to address former injustices by opening up opportunities to new markets for new entrants from previously disadvantaged population groups. The joint venture approach is intended to reduce the risk of failures due to the inexperience of such newly emerging companies and consultants by allying them to more experienced partners. The project organisation structure (figure 3.3) shows how this was done for the college IT centre project. While this arrangement does reduce the likelihood of occurrence for the technical risk of design failure, it actually increases the managerial risk of communication failure, since the project complexity is now increased through the higher level of project organisation differentiation – more people are engaged in doing the same tasks – and interdependency, since the task co-ordination requirements are also increased. The organisational structure diagrams for the architectural consultancy (figure 3.4) and the mechanical and electrical engineering consultancy (figure 3.5) illustrate the additional complexity involved.

In the architectural joint venture, the simplicity of the organisational structure for the affirmative action partner (JV1 Architect 'B' in figure 3.4) contrasts with the greater complexity of the lead partner (JV1 Architect 'A'). The latter organisation operates in a somewhat ad hoc manner and this, too, contributes towards potential communication complexity for the project. Architect 'B' in JV1, for example, might communicate a design decision to one of the three directors of JV1 project architect firm 'A' but, for one reason or another, this decision might not flow through to the site and be replaced by a different decision from another of the three directors. The potential outcome could be the technical risk of design failure.

By comparison, the organisational structures of the mechanical and electrical engineering consultancy JV2 partners are relatively simple, with clear communication paths.

To be fair, despite the added complexity introduced by the joint venture procurement arrangements in this project, it is likely that the benefits of mitigating the risk of design failure (due to inexperience) will outweigh the potential consequences of design failure risk due to inadequate communication.

Figure 3.6 represents the organisational structure for the prime stakeholder (the college) and reveals a further complexity in the communication paths from the highest level of decision-making to the project implementation level for this stakeholder. (We will explore the intricacies of organisational structure more fully in chapter 4, together with their implications for risk management.)

A more serious risk faced by the college example is the direct financial risk of failing to complete the project within the budget. The college has no previous experience in applying stringent cost management to its construction projects, despite the extensive built facilities already existing on the campus. All previous major building work was carried out under the direct supervision of a central government public works department, and the college itself had no

Figure 3.3 Organisational structure for college IT centre project

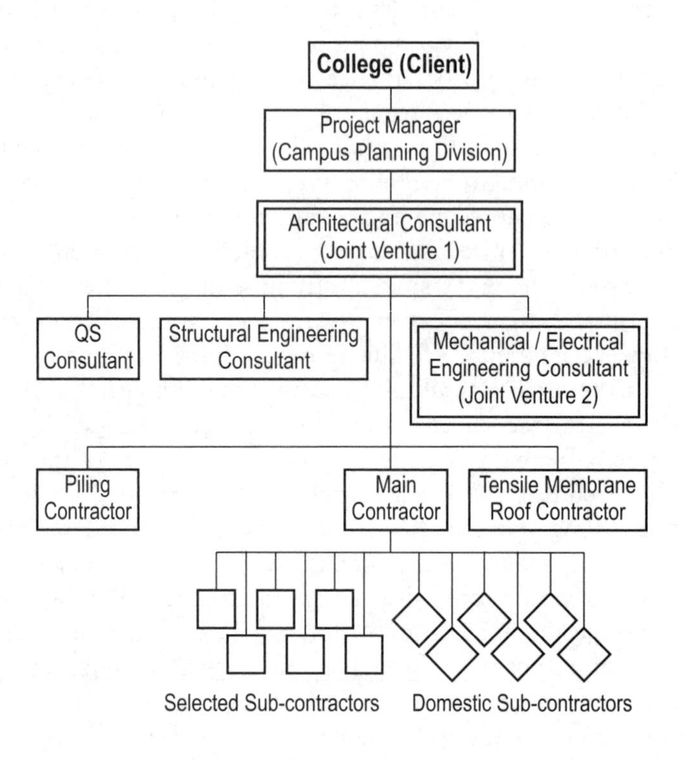

Figure 3.4 Organisational structure for architectural consultancy (JV1) for college IT centre project

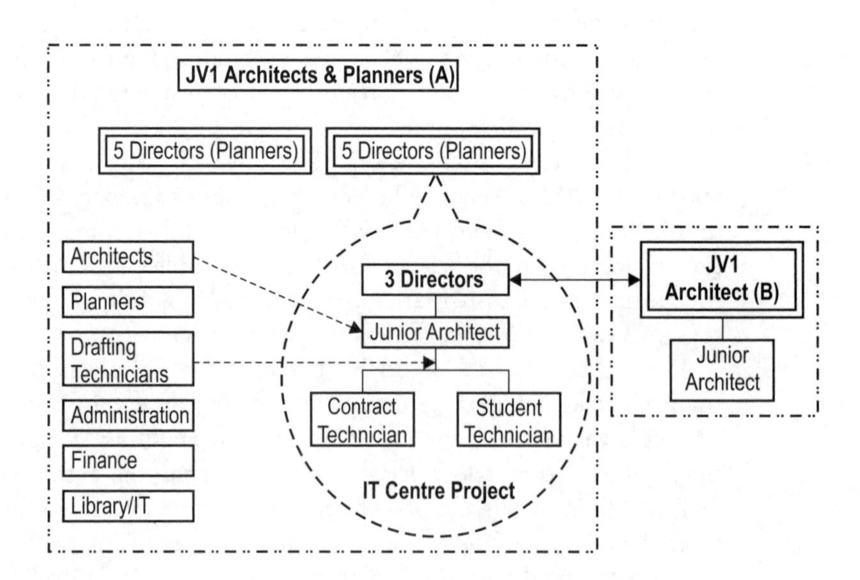

Figure 3.5 Organisational structure for mechanical and electrical engineering consultancy (JV2) for college IT centre project

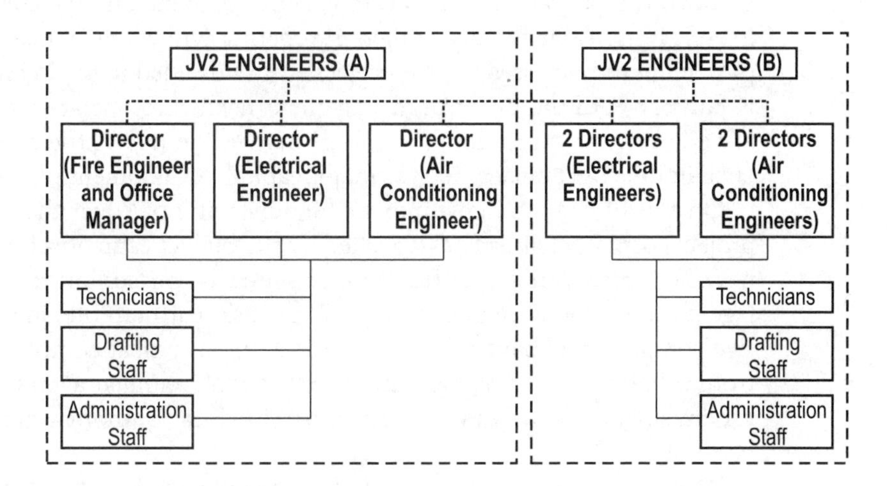

Figure 3.6 Organisational structure for the college client stakeholder for the IT centre project

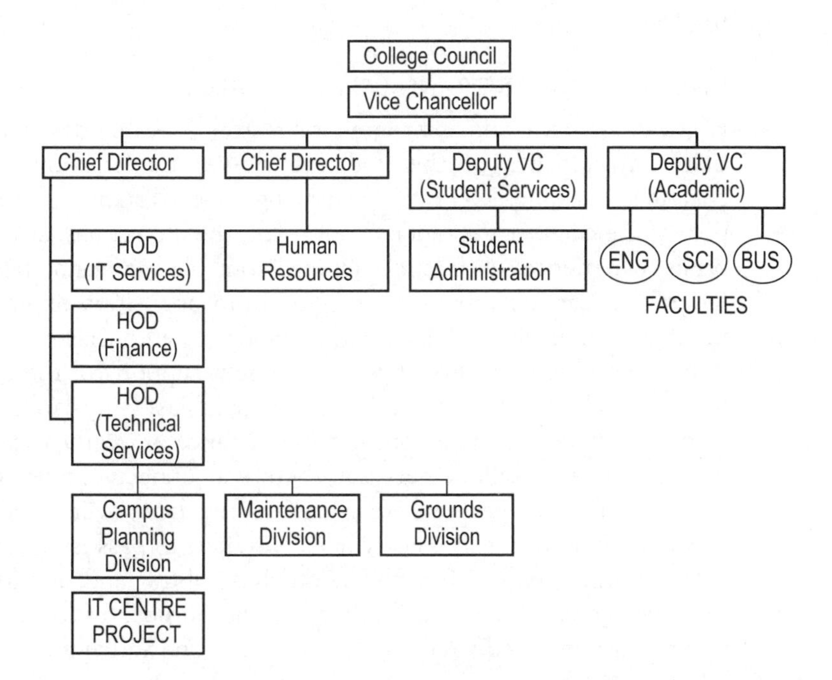

hands-on financial involvement. If previous projects overran their budgets, the government simply paid up.

Now, however, the government has provided a fixed amount towards the capital funding of campus development, and the college itself is responsible for allocating this, together with all the remaining procurement processes for the projects it has decided upon. While it is familiar with the financial management needed to operate a tertiary education organisation, the college management has little experience in the cost management of major capital development projects.

On top of this, the government funding will not cover the total project cost target already established by the budget estimates for the new IT centre. Additional funds will have to be raised, but the college has no experience or expertise in this task. Furthermore, innovative solutions will have to be found, since there is great competition from other fund-seeking institutions in the more familiar avenues of fundraising, such as industry sponsorship, local authority grants, bequests, etc.

What at first sight might seem to be a comparatively low risk project takes on a different complexion when the issues surrounding the IT centre are examined in greater detail. In terms of the factors discussed earlier, this is a somewhat risky project due to: (3) technology new to the user, (5) novel operational requirements by the client, (8) severe time constraints, (9) inexperience of some stakeholders, (10) stringent quality or performance standards, and (12) political and cultural sensitivities.

UNDERSEA POWER TRANSMISSION PROJECT

Political, technical and environmental factors probably dominate the risk characteristics of the undersea power transmission project described in example 3.3. While enjoying the consistent political support of the client stakeholder government, the project has, at varying times, experienced different reactions from the mainland state and federal governments involved, ranging from opposition on environmental grounds, through lukewarm support, to enthusiasm.

In Victoria, the level of state government support for the project has been influenced by the political climate in East Gippsland, where a converter station and a considerable distance of overhead powerline transmission will be necessary before a connecting point to the national grid can be reached. Much of the land over which the pylon-supported transmission cables will have to pass is prime agricultural land or natural forest. The purchase of the land, and its subsequent alienation and sterilisation for the purpose of locating the pylons, has provided a rallying point for a strong agricultural protest lobby.

Example 3.3

UNDERSEA POWER TRANSMISSION PROJECT

The Basslink Project is a high voltage electricity interconnector project to link the Tasmanian electricity grid into the national mainland grid in Australia. The project comprises converter stations in Tasmania and Victoria, and an undersea link approximately 250 km long laid on the seabed across the Bass Strait.

The client stakeholder comprises the state government of Tasmania and its hydro-electricity generation company. The client has signed a BOO (build-own-operate) agreement with the Australian subsidiary (formed expressly for the project) of a British organisation which is itself a subsidiary of a larger UK-based company operating in the field of major electrical distribution networks. The BOO concessionaire acts as the project developer, and has entered into an engineering, procurement and construction contract (similar to a design/ build contract) with a joint venture company formed between two European engineering firms, one German and the other Italian. Additional essential connection contracts have been negotiated between the BOO concessionaire and other power distribution and engineering companies in Victoria and Tasmania.

The aim of the project is to allow transmission of Tasmania's hydro-generated electricity to the Australian mainland, and thus take advantage of the island's surplus power capacity during the prevailing wet periods of its temperate climate. The link would also permit transmission in the opposite direction, should prolonged drought affect Tasmania's ability to generate sufficient electricity for its own needs.

The project development period is estimated to take up to 5 years, with construction requiring a further 3 years.

Concerned environmentalists have lobbied for a more expensive underground cable alternative on the mainland. Yet another lobby group has expressed fears about the unknown electro-magnetic radiation effects of the high voltage transmission lines (whether overhead or underground) on nearby communities and animals. Final transmission routes on the mainland have yet to be announced, and this has provided more fuel to protest groups in the NIMBY (not in my backyard) category.

Besides the political and environmental sensitivities of this project, there are obvious weather risks. Any undersea project is faced with this, but the Bass Strait is renowned for difficult weather conditions at various times. The litter of shipwrecks along the western seaboard of Victoria and off the coast of South Australia bears witness to this.

The technical risks are not insignificant, despite the fact that the technology intended for the project is not radically new, nor is it new to the stakeholders. The nature, scale, distance involved and overall time required for the project, however, mean that some of the existing technology could be overtaken by new developments in power transmission, or that control equipment may become obsolete, or that plant and equipment required for construction will prove inadequate.

The mixture of European and Australian companies involved in the project suggests that there will be some level of cultural sensitivity to deal with within the project organisation, as well as any encountered in the surrounding communities. Communication standards will have to be high.

Finally, the array of contractual agreements required between the various stakeholders, the compulsory purchase procedures, the wayleave arrangements for transmission cable routing, and the transmission tariff agreements, all contribute to a substantial legal complexity in this project. While the legal instruments involved may be subject to Australian contract law, the European domiciles of the two joint venture contractors and the holding company behind the main developer will be a complicating factor should disputes arise.

The examples described above illustrate some of the factors that can contribute to the riskiness of projects. An important point to observe is that in none of them are all the risk factors precisely quantifiable. This reinforces the point made at the beginning of this section; that riskiness is a relative, rather than absolute, characteristic of projects.

3.6 CHAPTER SUMMARY

This chapter should have given you a better understanding of the nature of projects as undertakings established to achieve predetermined objectives with a given time-frame. We have adopted a multi-environment view of projects, to encompass procurement, operational and disposal stages, although not all projects will include all of these.

The constituent elements of any project, regardless of type, comprise the tasks, technologies, resources and organisation required to undertake it. Each element is capable of further division into more detailed project-specific sub-elements.

The complexity of a project is influenced by the degree of differentiation, together with the nature and extent of interdependency, discernible in its elements and sub-elements.

The elements and sub-elements of a project involve a substantial amount of decision-making. Decision-making gives rise to risks, largely because of the uncertainties inherent in the variables that must be considered in the decision-making process. These uncertainties are exacerbated by the relative complexity of the project.

These and other factors contribute to the 'riskiness' of projects, and provide a rationale for the need to manage the risks involved in undertaking them. Before considering how project risks should be managed, however, it is first necessary to consider the organisational aspects of projects in terms of the different stakeholders involved. This will provide the necessary transparency to stakeholder decision-making that will allow us to explore risk management in the most appropriate way.

CHAPTER 4

PROJECT ORGANISATION

4.1 INTRODUCTION

According to Simon (1969, cited by Mintzberg, 1983), the essence of the man-made sciences – whether engineering, medicine, or management – is *design*. Design, by definition, assumes discretionary powers, an ability to alter a system. In the case of organisational structures in general, and project organisations in particular, design means influencing the division of labour and co-ordinating mechanisms, thereby affecting how the organisation functions. Part of this design process entails the design of the decision-making system, that part of the organisation responsible for the division of labour and systems of formal authority, regulated flows, work groupings, and ad hoc decision processes, as well as systems of informal communication, work groupings and ad hoc decision processes. One may ask why this material is relevant to a book on project risk management? If one considers that project risks arise, for the most part, from the decision-making of individuals working in organisations, then an understanding of organisations and their associated decision processes is important.

This chapter provides an overview of organisational theory, organisational structures, and project stakeholder organisations and structures. Emphasis is placed on the relationship between decision-making within project organisations and project risk.

4.2 ORGANISATIONAL THEORY

There is no one single theory of organisations. Rather, each of the many theories that have been documented provides a unique perspective of an organisation and thus any phenomena being studied within the organisation. Since risk is seen as a social construct (see chapter 2) the behaviourist school of management thinking is appropriate here.

The role of people in organisations encompasses the human relations and decision-making approaches to management thinking. Stemming from the work of Mayo (1933, 1949), Maslow (1943) and McGregor (1960), the behaviourist school of management conceived organisations as co-operative systems of interacting and interdependent interest groups which collectively shape the organisations' goals and how they are achieved. Groups and their individual members are themselves seen as having needs and contributing a set of beliefs and attitudes to the organisation to which they belong. This set of beliefs partly determines the behavioural responses of the organisation to various stimuli.

In terms of the behaviourist school of management thinking, the importance of both vertical and horizontal information flow is acknowledged, with the manager's task being seen as one of organising and sustaining co-operation and teamwork by satisfying people's needs. In turn, managerial effectiveness is measured by the degree to which responsibility is informally accepted (Loosemore, 1996). Two managerial advances set this philosophy apart from its predecessors. Firstly, risk and uncertainty are seen as a natural part of organisational life, the emphasis being on the acceptance of instability and the need for adaptability to accommodate risk and uncertainty rather than suppressing it. Secondly, there is recognition of the importance of and the role played by informal interest groups. This perspective of management is clearly of direct relevance to any study of project risk management.

The basic tenets of the human relations approach to management were extended by the works of writers such as Simon (1948), Selznick (1957) and Crozier (1964), who focused attention on the issue of power within organisations. According to Loosemore (1996), by initiating the so-called political science tradition, these theorists acknowledged the existence within organisations of differing goals, power tactics, manipulative behaviour, politics and conflict within and between different interest groups. For the first time, power struggles and conflict were seen as inevitable and legitimate within organisations. Such power struggles and conflict were seen to have both positive and negative implications depending upon the manner in which they were managed.

Notwithstanding the considerable body of knowledge on the subject of organisational conflict, attempts at deriving a precise definition of power have been difficult in socio-political theory. Suffice it to say here that, during periods of organisational instability and uncertainty, political 'savvy', an understanding of power and the ability to manage uncertainty and crises are essential for organisational survival (Sagan, 1994). Clearly, for the reasons given above and because of the conflicting interests that characterise project organisations, theories relating to power and conflict in a sociological and political sense are considered highly relevant to project risk management.

The second school of thought, relevant to the human behaviour paradigm, is the decision-making movement (see Simon, 1960; Cyert and March, 1963). In terms of this movement, the focus was on how people made decisions and managers were seen as decision-makers involved in setting goals, deciding upon means to ends, and adapting to external pressures or internal problems (risk and uncertainty) which threaten organisational performance.

In an attempt to account for uncertainty and the need for change, Simon (1960) introduced the concepts of 'bounded rationality' and 'satisficing'. In terms of this rationale, Simon claimed that the inherent uncertainty of the business environment results in inadequate information supply and thus prevents decision-makers foreseeing the cause and effect relations of their actions. Simon termed these constraints 'bounded rationality'. The tendency of decision-makers, under these constraints, to elect satisfactory rather than optimum solutions to problems was termed 'satisficing'. Simon went further, claiming that, decision-makers, often driven by self-interest, tend to restrict the range of alternative solutions they consider as a means of satisfying those interests.

The importance of the decision-making school of thought to any study of project risk management is transparent. Moreover, the principles of decision-making are also beneficial as they highlight the behavioural factors which can influence the effectiveness of a decision-making process. This is considered further in chapter 5.

The organisation as a system

The theoretical emphasis within organisational theory of the organisation as a system is represented by the systems and contingency schools of thought. Cleland and King (1983: 17) define a system as 'an organised or complex whole; an assemblage or combination of things or parts forming a complex or unitary whole'. The origins of systems theory are attributed to Von Bertalanffy (1969), a biologist who developed systems theory to explain the complex functioning of nature. According to Katz and Kahn (1978), the value of systems theory to the study of organisations lies in its concern for problems of structures, relationships and interdependence. In terms of systems theory the organisation is seen as a combination of interdependent parts or sub-systems which collectively constitute the whole or whole. Thought of in biological terms, the organisation is conceived as a living organism which has to adapt continually and naturally to changes in its environment in order to survive. Environment in this sense means both an internal and external environment.

Systems have been categorised as either closed or open (Von Bertalanffy, 1969). In terms of a closed-system, an organisation is seen as closed in the sense that it is independent of its environment; an open-system emphasises the interdependence between the organisation and its environment. The value of adopting the open-systems perspective for a study of project risk management is that it provides an understanding of why organisations and projects face risk and uncertainty within their business environment and provides a vehicle for understanding organisational response to that risk and uncertainty.

Systems theory has not been without its critics, with Mills (1967) cautioning that the adoption of a systemic view of organisation should not ignore the way that human emotions, needs, aspirations and behaviour permeate human work. More specifically, Mills stresses that the issue of management control is essentially a problem of human behaviour. Silverman (1970) warns that a view of the organisation as an independent living entity capable of its own actions is incorrect; rather, organisations do not (indeed cannot) react to their environment, the members of the organisation do.

Contingency theory, itself stemming from a refinement of, the open-systems perspective, was developed by theorists such as Burns and Stalker (1961), Woodward (1965), Thomsen (1967) and Lawrence and Lorsch (1967). In essence, their work explored more comprehensively the relationship between organisations and their environments and concluded that there was no one best way to organise for all situations. This finding was in stark contrast to all theories that had gone before, which had sought to establish the optimum solution to all organisational problems – the one best organisational structure! In terms of this approach to organisational design, the goal of managers should be to maximise the degree of fit between an organisation's internal structure and its environment. Problems, in the form of imbalance, can then only stem from two sources, namely, organisational (internal) and environmental (external). Clearly, such imbalance can be remedied by changing one or both of these components. This notion of fit led to efforts at describing the structures which best suited different environmental conditions, for example, organic versus mechanistic forms of organisation (Burns and Stalker, 1961). Organic structures typically incorporate informal authority structures, free vertical and horizontal information flow, attitude flexibility and high task commitment. In contrast, mechanistic structures emphasise vertical information flow, rule-bound behaviour, formal authority, and a lack of task commitment (Burns and Stalker, 1961).

Contingency theory has also had its detractors. For example, highlighted weaknesses include structural change lagging environmental change (Stopford and Wells, 1972); confusion over the exact relationship between structural and environmental factors; the level of abstraction of organisational concepts and their metrics (Mintzberg, 1979); and the degree of influence organisations can have over their environments (Booth, 1993).

Clearly, the value of the contingency approach to an understanding of project risk management process lies in the fact that it effectively reduces a complex, multi-dimensional problem to a two-component model, which consists of the environment and the project organisation. The theory is also valuable as certain organisational characteristics are more appropriate than others when the organisation is faced with certain environmental conditions (Loosemore, 1996). Equally, there are different communication and decision-making patterns in different risk management situations. Knowledge on the part of risk managers of such relationships would be invaluable in facilitating better risk judgements when faced with risky situations.

4.3 ORGANISATIONAL STRUCTURES

The structure of an organisation refers to issues such as the degree and type of horizontal differentiation, vertical differentiation, mechanisms for co-ordination and control, formalisation and centralisation of power and decision-making. Mintzberg (1983: 2) provides a succinct definition of the structure of an organisation: 'the sum total of the ways in which its labour is divided into distinct tasks and then its coordination is achieved amongst these tasks'. An organisation can be thought of as a system, itself belonging to larger systems and comprising sub-systems.

Mintzberg argues that the elements of structure should be selected to achieve internal consistency or harmony, as well as a basic consistency with the organisation's situation (size, age, environment, technical systems, etc.). The grouping of positions within an organisation is a fundamental means to co-ordinate work within the organisation. The effects of grouping include: establishing a system of common supervision among positions; sharing common resources including knowledge; creating common measures of performance; and encouraging mutual adjustment through informal communication.

According to Mintzberg (1983), the six bases for the grouping of positions commonly considered in the literature (i.e. grouping by knowledge and skill, by work process and function, by time, by output, by client, and by place) can be re-grouped into two essential categories. These categories are: market grouping, comprising the bases of output, client and place; and functional grouping, comprising the bases of knowledge, skill, work process and function, with time arguably falling into either category.

4.4 PROJECT ORGANISATIONS

A project organisation is complex in the sense that it possesses both intra- and inter-organisational interactions. Intra-organisational interactions and motivation can be explained by the relevant management theories on structure, motivation, communication, etc. However, such theories are inadequate to explain inter-organisational interaction. Project-based activity needs to be performed by multiple parties. Each party is a separate organisational entity possessing its own interests and anticipated rewards at project completion. By definition, the project is a temporary, inter-organisational venture that lasts only for the duration of the project (Dulaimi, Ling and Bajrachary, 2003).

Clearly, inter-organisational co-ordination of these various organisations is difficult, rendering innovation initiation and implementation difficult (Dulaimi, Ling and Bajrachary, 2003). The success of a project depends in no small measure on the existence between the various parties of favourable and mutually supportive relationships. Moreover, co-ordinated information and resource exchange between the parties must also exist.

The transient nature of projects and the associated coming together of team members, often at different points in time and for different durations, may not motivate members to propose, initiate and implement innovation on that particular project (Dulaimi, Ling and Bajrachary, 2003). Winch (1998) highlights the possible negative aspects of this fragmentation, claiming that team members are more likely to be concerned with completing the project to fulfil their individual interests than with innovation. As Borys and Jemison (1989) note, this fragmentation may be caused by a lack of understanding of the common project objectives and each other's individual interests, resulting in each team member acting on the basis of individual rationality. This, in turn, is likely to give rise to conflicting tension among team members. Bowen et al. (2000), researching the group dynamics of project teams within a construction project delivery context, found not insignificant evidence of shortcomings on the part of team members regarding: role responsibility definition; role ambiguity; understanding of client goals; adoption of client goals; and team conflict.

According to Knoke (2000, cited by Dulaimi, Ling and Bajrachary, 2003), the motive behind organisational relationships can be explained by three theories: institutional; transaction cost; and dependency theories. In terms of the institutional theory, an organisation seeks alliance with other suitable ones because of a normative institutional requirement to maintain socio-political legitimacy in the field. In terms of transaction cost theory, an alliance is formed if one organisation decides to purchase a particular function from another organisation. The dependency theory postulates that an organisation establishes a relationship with another organisation to acquire a specific function from the other organisation because of a deficiency in that resource.

The management of projects necessitates the arrangement of groupings of the requisite skills, resources and knowledge required to undertake the project. This is true whether the project is being undertaken in-house, or if it is being undertaken by a number of persons drawn from a number of parent organisations. For the sake of simplicity, we can term such specially formed groupings as 'project organisations'. Of particular relevance to project organisations is grouping by knowledge and skill, where typically persons from different disciplines bring disparate skills together to achieve the project goal(s). The issue of project organisational structure is complicated by the fact that project team members themselves typically belong to 'parent' organisations. Each parent organisation will possess an organisational structure commensurate with its core business activities and the nature of the environment within which it operates. For example, the stakeholders that typically comprise a construction project team (e.g. project manager, designer, engineer, cost engineer/ consultant, and contractor) are themselves members of organisations whose organisational structures may each be quite different. Such project management teams may be considered to be temporary management structures in that such teams form at the commencement of a project and disband upon its completion. Membership of the project organisation is thus transient, while project team individuals' membership of their parent organisations is typically for much longer periods. These project teams are goal directed and are called 'matrix organisations'.

Matrix organisations

A matrix organisation has been defined as a vertical functional hierarchy overlain by lateral authority, influence or communication, i.e. a mixed organisation (Knight, 1976). Mintzberg (1983) states that by using a matrix structure, the organisation avoids using one basis of grouping over another; instead, it chooses both. According to Rowlinson (2001), having such structures can lead to institutionalised conflict which, if properly directed, can lead to a number of advantages such as the efficient and flexible use of resources, technical excellence of solutions, and motivation and development of members. Matrix organisations are claimed to improve efficiency and effectiveness in a time of instability, resource constraints, and tight project deadlines. Rowlinson (2001, citing Handy, 1985) states that matrix organisations are most effective only when a particular organisational culture takes hold in the organisation, i.e. there is a tendency to see the task as the key issue in the organisation and to adopt, flexibly, whatever means considered to be appropriate for accomplishing the task. Matrix organisations are not without their problems. Mintzberg (1983) reports that, while most effective for developing new activities (projects) and co-ordinating complex multiple interdependencies, matrix structures do not provide for stability and security. Moreover, dispensing with the principle of unity of command often creates considerable confusion, stress and conflict, and requires from its participants highly developed interpersonal skills and considerable tolerance for ambiguity. Knight (1976) draws attention to the need, within matrix organisation, for more communicating to be done and for more information to be shared between participants. Coupled with these ills are the problems of maintaining the delicate balance of power between the different sorts of project team members, and of power distance (Rowlinson and Root, 1996).

In a typical project matrix organisation there is always some degree of tension between the needs of the project and the needs of the functional organisations to which members of the matrix organisation belong. There also exists the problem associated with the dual reporting by team members to both project and functional managers. This tension needs to be negotiated until a balance is obtained. A matrix organisation emphasises decentralisation of decision-making and relies on individual members taking responsibility for task production (Rowlinson, 2001). This necessarily entails a high degree of commitment to the organisation's values, normative commitment, and the acceptance of professional responsibility. The organisational structures of individual members' parent organisations can also influence the effectiveness of the project matrix organisation. Rowlinson (2001) found, for example, in a study of the effectiveness of a transition to matrix organisations in the Hong Kong public sector, that a mismatch between organisational structure, procedures and culture was a key problem.

4.5 STAKEHOLDER ORGANISATIONS

Freeman (1984: 46) defines a stakeholder as: 'any individual or group who can affect or is affected by actions, decisions, policies, practices, or goals of the organization'. Stakeholders are generally classified as primary or secondary. According to Cleland (1998), primary stakeholders have a contractual or legal obligation to the project team. In addition, they have the responsibility and authority to manage and commit resources according to schedule, cost, and technical performance objectives. Examples of such primary stakeholders would include: the client, the project team, the consultant organisations, finance organisations, contractors, and sub-contractors. Secondary stakeholders typically comprise all other interested groups, such as the government, local authorities, media, consumers, competitors, public and society.

There has been a growing trend toward recognising a greater participation of society with an interest or 'stake' in projects and organisations. As an organisation's success can be affected negatively or positively by relationships with its stakeholders, the business requires careful management attention in considering the demands of its stakeholders (Post et al., 1996). Stakeholder theory has been advanced and justified in the management literature on the basis of its descriptive accuracy, instrumental power and normative validity (Donaldson and Preston, 1995). Gibson (2000) summarises the difference between these three approaches, stating that the descriptive approach tells whether stakeholder interests are taken into account, whereas the instrumental approach is concerned with the impact stakeholders may have in terms of corporate effectiveness. The normative approach deals with the reasons why corporations ought to considered stakeholder interests even in the absence of any apparent benefit.

In a project management context, primary stakeholders are seen as participants who are directly involved in the project or have contractual agreements. These primary stakeholders include clients, contractors, suppliers, investors and designers. In essence, these are the members of the project management team and can be thought of as 'internal primary stakeholders'. The situation is complicated as these internal primary stakeholders themselves belong to stakeholder groups in the form of the parent organisations to which they belong. Consequently, the internal primary stakeholders may themselves be thought of as comprising direct and indirect internal primary stakeholders.

In a project context, the secondary stakeholders are individuals, groups and organisations who are not directly related to the core business of either the project team or the organisations to which the project team members belong. These secondary or external stakeholder groups include government, local authority, local communities, and consumer groups (Preece, Modley and Smith, 1998). The secondary stakeholders can exert a significant influence on the development of the project; particularly government which can exert influence through the use of legislation. Since construction projects invariably have some sort of impact on the surrounding environment, construction project teams have often had an adversarial relationship with environmental groups and local communities. They can take action in the form of active lobbying, or more direct action targeted at the construction process in an attempt to change the plans or construction activities.

Organisation/stakeholder relations can change over time. Generally a realignment could occur if: institutional support changes; contingent factors emerge; ideas held by stakeholders and/or organisations change; or where material interests on either side change. Friedman and Miles (2002) provide an example of such realignment in their discussion of Greenpeace moving from being antagonistic and favouring violent confrontation, to a position of providing solution-based approaches to problems and forging corporate alliances with their former opposition. Examples of such collaboration include Greenpeace and British Petroleum Solar collaborating on ways of developing solar energy and Greenpeace and UK retailers such as Tesco working together on the impact of PVC packaging and building materials (Friedman and Miles, 2002).

4.6 STAKEHOLDER MANAGEMENT

Project stakeholder management aims to reduce uncertainties from stakeholders' activities that might adversely affect the project and to encourage stakeholder support of project goals or objectives. It adopts the concept of instrumental stakeholder theory (Cleland, 1998).

The instrumental perspective of stakeholder management believes that stakeholder management activity can lead to better outcomes, e.g. higher profitability, increased firm value, improved predictability of changes in the external environment, fewer incidents of managing adverse moves by stakeholders (e.g., strikes, boycotts) (Harrison and St John, 1996). For example, Kotler and Heskett (1992) found a strong tendency for companies that considered the interests of all major stakeholder groups in their decision-making to achieve better performance. The major benefit of proactive stakeholder management is perhaps to create organisational flexibility in response to environmental change, resulting in quicker response times (and hence potentially lower costs) to environmental impacts.

The instrumental approach emphasises three process, namely: identifying stakeholder attributes; determining stakeholder interests; and assessing dimensions of their power and influence (Frooman, 1999). Frooman points to recent trends whereby organisations not only concentrate on identifying stakeholders and determining the types of influence they exert on the organisation, but also develop response strategies to those influences.

To summarise, organisations interpret and give meaning to the organisation by seeing organisational practices, actions and policies reflected in the explicit and implicit behaviour of stakeholders (Scott and Lane, 2000). The perceptions of managers of the impressions that stakeholders form of an organisation are termed reflected stakeholder appraisals. In this regard, Scott and Lane (2000: 53) proposed that: 'managers will engage in active processing of an organization's identity-relevant information when reflected stakeholder appraisals are negative and when they threaten important aspects of managers' self-identities and perceived organizational goal commitment'. Moreover, the more dense the stakeholder network, the greater the influence of stakeholder needs, values and beliefs will be on organisational managers. Finally, the greater the extent to which an organisation perceives a stakeholder to be powerful, legitimate, and having an urgent claim, the more they will consider the stakeholder's needs, values and beliefs.

4.7 ORGANISATIONAL BEHAVIOUR AND ORGANISATIONAL LIFE-CYCLE

Jawaahar and McLaughlin (2001) integrate organisational life-cycle theory, resource dependency theory, and stakeholder management theory. In terms of their descriptive stakeholder theory, they: show that at any given organisational life-cycle stage, certain stakeholders, because of their potential to satisfy critical organisational needs, will be more important than others; identify specific stakeholders likely to become more or less important as an organisation evolves from one stage to the next; and propose that the strategy an organisation uses to deal with each stakeholder will depend on the importance of that stakeholder to the organisation relative to the other stakeholders. It must be appreciated, however, that organisations do not respond to each stakeholder individually; rather, they respond to the multiple influences of the entire stakeholder set (Rowley, 1997).

Because threats and opportunities vary with life-cycle stages, organisations are likely to have different needs, in terms of resources, in different stages of the organisational life-cycle. Consequently, the relative importance of stakeholders will also change as the organisation evolves through the stages of start-up, emerging growth, maturity, and decline/transition.

The start-up stage of the organisation or project involves developing and implementing the business or project plan, financing, assembling the requisite resources, and (critically) client or customer acceptance. Jawahar and McLaughlin (2001) state that during this phase resource allocation decisions are framed in the context of losses, and a risk-seeking strategy of defence or reaction will be pursued to deal with those stakeholders not critical for organisational or project survival. More specifically, an organisation is likely to assume the risk of using the strategy of reaction to deal with trade associations and environmental groups, and the strategy of defence with government and community stakeholders. However, the organisation will use the strategy of proaction to deal with shareholders, credits and customers, and the strategy of accommodation to deal with employees and suppliers as these stakeholders are critical to survival.

During the emerging growth stage, Jawaahar and McLaughlin (2001) say that resource allocation decisions will be framed in the context of gains, and a risk-averse strategy will be adopted to deal with stakeholders. Issues of creditors, employees, suppliers and trade associations will be addressed proactively, whereas concerns of shareholders, customers and clients, government, communities and environmental groups will be accommodated. In the mature stage, resource allocation decisions will be framed in the context of gains, and a proactive risk-averse strategy will be used to deal with all stakeholders, except for creditors, who will be accommodated.

Finally, in the decline/maturity phase of the life-cycle, resource decisions will be framed in the context of losses, and risk-seeking strategies of reaction and defence will be used to deal with stakeholders not critical to organisational survival. Reaction will be used for trade associations and environmental groups, and defence for the government and the community. A strategy of proaction will be used for shareholders, creditors, and clients/customers, while accommodation will be used for employees and suppliers, since these stakeholders are critical for survival.

4.8 CHAPTER SUMMARY

This chapter has provided an overview of organisational theory, organisational structures, and project stakeholder organisations and structures. It is clear that organisations exist within the context of both internal and external environments. An understanding of organisational theory is useful as certain organisational characteristics are more appropriate than others when the organisation is faced with particular environmental conditions and changes. Equally, there are different communication and decision-making patterns in different risk management situations. Knowledge on the part of risk managers of such relationships is useful in facilitating better risk judgments when faced with risky situations.

Project organisations may be thought of as temporary, inter-organisational structures. As such, project organisations possess both intra- and inter-organisational interactions, rendering them complex from a management point of view. The effectiveness of project organisations is further complicated by the presence of stakeholder organisations. The challenge facing project organisations is the management of their relationships with these stakeholder organisations within a turbulent and risky environment.

In the next chapter we will consider the decision-making and communication environments of organisations, as well as their implications for risk perceptions and risk message communication.

CHAPTER 5

DECISION-MAKING, RISK PERCEPTION AND RISK COMMUNICATION

5.1 INTRODUCTION

In this chapter decision-making, risk perceptions and communication are discussed within the context of organisations and project teams. The chapter commences with an examination of the decision-making and communication environment of organisations, and then explores these in terms of risk perception and risk communication.

5.2 THE DECISION-MAKING ENVIRONMENT OF THE ORGANISATION

According to Simons and Thompson (1998), decision-making comprises the act of seeking information, interpreting that information, and, based on such perceptions, arriving at a conclusion.

Ansoff (1965) identified three categories of decisions, namely, strategic, administrative and operating decisions. Strategic decisions are concerned with the external rather than the internal environment of the organisation, involving considerations such as the organisation's objectives and goals. Administrative decisions are concerned with structuring the organisation's resources to maximise the organisation's performance potential. They would include issues such as the nature of the organisational structure, information flows, and communication patterns. Operating decisions are aimed at maximising profitability of operations, and are the focal point of the endeavours of most organisations. These categories are essentially derived from the functional divisions of the organisation, and are typical of the 'management approach' (Skitmore, 1986). Fellows et al. (1983) provide another classification of decisions, the distinction being between 'strategic' and 'tactical' decisions. The former focus on 'what shall we do?', whereas the latter address the question of 'how shall we do it?'.

According to Simons and Thompson (1998), four basic themes are evident in research into nature of managerial decision-making: environmental factors; organisational factors; decision-specific factors; and individual managerial characteristics and values.

Rajagopalan, Raheed and Datta (1993) modelled the first three of these themes. The trend in studying environmental factors external to the organisation has focused largely on national culture, national economic conditions, and industry conditions. Research is this area stresses the importance to the organisation of accurate interpretation and prediction of environmental economic factors (Rajagopalan, Raheed and Datta, 1993). The implication of this is that the survival of the organisation will be in jeopardy if it is unable to predict, or at least respond to, the environment in which it exists, i.e. there will be a mismatch between its services and client demand.

Research into organisational factors has focused primarily on organisational structure, organisational culture, the structure of decision-making bodies, the impact of upward influence, and employee involvement. From this developed the school of thought that decision-making activities are not single events, but incremental, and that organisations are learning entities (Simon, 1973).

Investigation into decision-specific factors has focused largely on time, risk, complexity and politics (Butler et al., 1991) and, more recently, the dual explanations of strategic decision-making content, based on dimensions of complexity and politicality. Complexity relates to the multidimensionality of decision content and politicality refers to the divergence of stakeholder interest. Organisations process strategic issues differently depending on the importance, or criticality, of the issue (Dutton, 1986). In other words, it is quite possible that the type of decision-making process used could be related to the nature of the issue under examination.

The fourth theme has involved the study of individual managerial characteristics and values. Hambrick and Mason (1984) drew attention to the tendency in much of the literature to dissociate the process of decision-making from the people making the decisions. They proposed a model of strategic choice that related manager choice to observable demographic characteristics, arguing that such characteristics are indicators of values and cognitive abilities.

Simons and Thompson (1998) warn of the limitations of the above research, claiming that few studies incorporated a range of managerial characteristics, and suggest that the lack of sensitivity to the multidimensional interaction between environmental, organisational, and individual factors impinging on the decision-making process has led to results that are situation specific and difficult to generalise.

Attempting to address these limitations, Simons and Thompson (1998) focused their attention on whether or not managerial decision-making is influenced simultaneously by environmental, organisational, decision-specific, and individual factors. They found that private sector managers describe the impact of their decisions in financial and resourcing terms such as profit and market share. Public sector managers, on the other hand, described the impact of their decisions in structural terms such as workflow patterns and staffing efficiency. Managers defined decision-making in three major ways, namely: problem-focused, goal-focused, and political appeasement. Problem-focused definitions involved the identification of a problem, processing of information, decision-making, and delegation of tasks or duties aimed at resolving the problem. This view of decision-making is essentially a reactive process. Goal-focused definitions mostly involve the setting of goals or plans for organisational growth, ascertaining the feasibility of long-term plans, obtaining progress feedback on the attainment of those goals, and delegating long-term tasks. As such, goal-focused decision-making is essentially a planning process. The political appeasement definition of decision-making involved the need to satisfy conflicting stakeholders, identify divergent points of view, interpersonal skills, and encourage commitment to maintaining a constant work flow.

According to Simons and Thompson (1998), the definitions of problem- and goal-focused decision-making are very similar to the theoretical distinctions between operational and strategic decision-making. Their research supported the theoretical position of Rajagopalan, Raheed and Datta (1993) that decision-making is influenced simultaneously by a number of environmental, organisational, individual, and content-based factors – the greatest variety of factors falling within the environmental and organisational categories.

A simplified decision-making process

Skitmore (1986) provides a useful (albeit simplified) description of the decision-making process as comprising problem identification and selecting between alternative options.

PROBLEM IDENTIFICATION

Prior to the actual decision process, the decision-maker must be aware that a situation has arisen that requires a course of action to bring about a particular, desired state of affairs. This situation may be thought of as a problem-identification process, even though the issue at hand may not amount to a problem per se. Once the extent of the 'problem area' is known, the decision-maker is in a position to implement the actual decision process.

THE DECISION PROCESS

The decision-making process is essentially a question of selecting from a set of options. Theoretically, at least, an infinite set of options exists in all situations, but it is essential to reduce the solution space to more manageable proportions. It is necessary, in addition, as an aid to the selection procedure, to perform some evaluation of each option. This three-stage process of decision-making is called the decision choice process (Johnson and Scholes, 1984).

In identifying, evaluating and selecting decision options, it is necessary to consider the interaction between decisions and the environment (open social system) over time. Murray (1980) concludes that a decision system will need to recognise cultural, political and social inputs in an open system. In terms of a decision system, this implies the existence of several option-selection procedures that permit the decision-maker to evaluate the consequence of decision strategies, and to choose the strategy which satisfies personalised rationality (Wagner, 1971).

Identification of options

The identification of promising options is often a function of the evaluation and selection process. A factor influencing the inclusion of a potential option is the quality and quantity of the information needed and available, and the associated time and cost. Time and cost constrain the number of options that can be identified and evaluated.

Evaluation of options

Evaluation of each option implies that some knowledge is available about the future outcome of the decision option. As in the case of the option identification problem, the extent of this knowledge will depend on the quality and quantity of information available, and the associated time and cost factors. The more information possessed by the decision-maker, the better placed that individual is to make the decision.

The accuracy of the evaluation will depend on the ability of the evaluator. Further, the outcome of a decision is not necessarily independent of the decision-maker, who may participate in the implementation process (Skitmore, 1986). To facilitate objective selection procedures, each option needs to be evaluated in a similar manner, which implies the presence of some classification criteria.

Selection of an option

Difficulties in selection occur in accommodating conflicting criteria, particularly those evaluated qualitatively as opposed to quantitatively. Once again, information, cost, time and the ability of the selector are important issues.

Booth (1981) questions the necessity of identifying all the options prior to evaluation, and of evaluating each option prior to selection. According to Booth (1981), the evaluation of options is normally done as they are identified, as a result of the time and cost constraints. This approach to the decision process is representative of an iterative model of decision-making, where each option is identified in turn, evaluated, and compared with the previous best solution. In this manner the decision-maker has control over the width and depth of the decision process.

5.3 THE COMMUNICATION ENVIRONMENT OF THE ORGANISATION

Communication may be thought of as the 'social glue' that ties members of organisations, project teams, and other organisational sub-units together.

Definitions of communication

There are many definitions of communication. For example, Berelson and Steiner (1964) define communication as the transmission of information, emotions, and skills by the use of symbols – words, figures, graphs, etc. Dance (1967) defines communication within the broad framework of behaviourist psychology, as 'the eliciting of a response through (verbal) symbols'. Communication has also been defined as 'the mechanism by which power is exerted' (Schachter, 1951). Fotheringham (1966) views communication in terms of 'good or bad', 'effective or ineffective', and as directed at assisting a receiver to perceive a meaning similar to that in the mind of the communicator. Still other definitions (e.g. Newman, 1948) view communication in terms of a social aspect, namely, 'the process by which an aggregation of men is changed into a functioning group'. Clearly, interpersonal communication is a process by which meaning between sender and receiver is exchanged. By extension, effective communication is achieved when sender and receiver attain shared meaning.

Perspectives on communication

Applying perspectives specifically to the process of human communication raises issues which focus on the nature of communication. More specifically, it provides an indication of the points of analysis that can be distinguished among the various perspectives. The perspectives proposed by Fisher (1978) are classified as mechanistic, psychological, inter-actional, and pragmatic. The mechanistic perspective is adopted here for illustrative purposes.

Viewing communication within a mechanistic perspective implies a form of conveyance or transportation across space. The components of this perspective consist of the message (travelling across space from one point to another), the channel (the mode of conveyance of the message), the source and receiver, encoding and decoding (the process of transforming a message from one form to another at the point of transmission and destination), noise (the extent to which the fidelity of the message is reduced), and feedback (a message that is a response to another message).

Central to the mechanistic perspective of human communication is the element of transmission – the movement of the message by means of an appropriate channel. This channel, linking source and receiver, is clearly directional; the directional aspect implies impact on the receiving end, fostering the notion of the source influencing the receiver, i.e. if (specified message variables), then (receiver effects). The transformation of the message via the encoding/decoding process is highly complex, involving linguistic codes, paralinguistic cues, learned behaviours, cognitions and sociocultural norms (Fisher, 1978). Barriers to communication are seen mainly as existing within the individual's limited capacity to process information received from multiple sources, as opposed to the perception that barriers stem mainly from psychological restrictions inherent in the individual's cognitive capacity for encoding and decoding messages.

Within this perspective of human communication, the meaning of a message is seen as being a function of the location of the message at some specific point along the process of communication – in other words, meaning prior to encoding, meaning encoded and transmitted, meaning received and meaning decoded. The attractiveness of the mechanistic perspective lies in its simplicity and its emphasis on the physical components of communication.

Models of communication

Various models have been developed in attempts to explain the communication process. Examples of such models include those developed by Lasswell (1948), Shannon and Weaver (1949), Johnson

(1951), Newcomb (1953), Schramm (1955), Gerbner (1956), Westley and MacLean (1957), Berlo (1960), Ross (1965), Feldberg (1975) and Tubbs and Moss (1981). These models all view communication from different standpoints and, consequently, can be loosely categorised into one or another of the perspectives of human communication mentioned above.

The communication model of Feldberg (1975) is linear and mechanistic in nature, and may be described in terms of four distinct stages. The first stage is one in which a simple two-person communication process is assumed. The relevant components at this stage are the original sender, the final receiver, the message, the medium, the expectations of the sender, the reactions of the receiver, the result or outcome, the direction of the message and the content of the message.

The sender communicates the message to the receiver with some objective in mind. This objective may be conscious or unconscious, structured or unstructured, and is translated into the sender's expectations. The message is transmitted to and received by the final receiver, who reacts in some way to the message. The sender's expectation (anticipated result) and receiver's reaction (actual result) are translated into a result. Whether the receiver's reaction conforms to the sender's expectation depends on many factors, the aim of the communication process being to ensure that the sender's expectation and receiver's reaction are as congruent as possible. Prior to transmission, the sender should know the identity of the receiver if the correct expectations are to be established.

The degree of congruence is affected by the direction, medium and content of the message. The direction of the message refers to the route taken by the sender in transmitting the message, while the medium refers to the means used to transmit the message. The sender should endeavour to select the most effective method for transmission, the choice of medium depending on the availability of media, the content of the message, the nature of the receiver, and the distance between sender and receiver. The message transmitted by the sender will possess content or information. The exact content will be influenced by the direction of the message, the nature of the receiver, the nature of the medium, and by the objectives of the sender.

The second stage involves examining some of the major reasons why the sender's expectations and receiver's reactions are incongruent. The principal concepts here are personal factors, external pressures, physical noise, psychological noise, and personal defence mechanisms. Personal factors include such influences as age, gender, status, profession and value system. Differences in these personal factors cause different perceptions of reality. External pressure refers to those pressures emanating from other individuals or groups that may cause incongruent expectations and reaction notwithstanding similar perceptions of the message (e.g. lobby groups reacting to a controversial project). Physical noise is self-explanatory. Psychological noise and personal defence mechanisms are similar in nature. The former refers to noise caused by fear, anxiety and insecurity on the part of the receiver (and here stakeholder risk perceptions are seen as influential). Psychological noise can stem from fear of the sender and/or the contents of the message, and serves to distort the communication process. The latter form of noise refers to the fact that many of the receiver's personal defences are based on deep-seated experiences and values, many of which may be based on the relationship with the sender or on prejudices. It is these experiences and values that influence stakeholders' attitudes towards, and responses to, project risk.

Stage three constitutes the mechanism for evaluating the relative success of the communication process. Evaluation takes place on the part of the sender, the evaluation being in the form of feedback. Feedback on the relative congruence between expectation/reaction and anticipated result/actual result derives from three main sources. Firstly, feedback emanates from the receiver. The receiver may be required to, or may of the individual's own volition, inform the sender of the result of the reaction to the message. Secondly, feedback may come from the sender. The sender may check on the receiver's reaction, and the result of this reaction. Lastly, feedback may derive from other people or groups. Other people may have an interest in the receiver's reaction, reporting their evaluation of this result to the sender. In all three cases, the senders evaluate the validity of the feedback in terms of its source and in terms of their own expectations.

The final stage of communication revolves around the notion of two-way communication. Immediately the receiver responds to the sender's message with a response message, the communication process is again initiated, but in reverse. In essence, all the factors mentioned in the previous three stages are brought into play, being now applicable to the receiver as sender and the sender as receiver. In all likelihood the original sender will respond to the original receiver's message, and the entire cycle of communication will begin afresh and continue until the communication process is terminated. An example here might be the allocation of a risk to a project stakeholder; that risk taker's response in the form of a bid price; and a subsequent process of price negotiation.

In its simplest form, the mechanistic model assumes direct communication between sender and receiver. In any organisation a substantial amount of communication passes through, and is interpreted by, several intermediate receivers. In a project, this might be exemplified by the nature of client–contractor–subcontractor relationships. This diminishes the likelihood of the final receiver's reaction being congruent with the original sender's expectation. According to Feldberg (1975), the main responsibility for ensuring that expectations and reactions are congruent lies with the sender or originator of the message. To facilitate the attainment of congruence, it is the sender's responsibility to ensure that the appropriate receiver is selected, the most appropriate medium is chosen, the message contains the correct content, and that there is an absence of psychological barriers inhibiting the ability of the receiver to accept the message.

Tubbs and Moss (1981) developed a model of human communication which addresses the static inadequacies of the Feldberg model, introducing the concept that communication transactions are dynamic. Their model portrays the movement of communication in time, and emphasises the mutually influential nature of the communication event. Dance (1967) observed that communication, while moving forward, is at the same moment coming back on itself and being affected by past behaviour. Communication is seen as ever-changing (dynamic), requiring the active participation of both sender and receiver. It is also perceived as convergent in that the source and receiver work together over time to create and share meaning – they converge on shared meaning.

Information, taken in its broader sense and pertaining more to knowledge than to data per se, can only be transferred between parties if that information is transmitted (and received) in a manner which is meaningful to both. Clearly, the more closely the models of reality of the transmitter and the receiver correspond, the more effective becomes the resultant communication. Thus, the effectiveness of the organisation and any project team is a function of the effectiveness of the communication within that organisation or project team.

5.4 PROJECT TEAM DECISION-MAKING AND COMMUNICATION

Project team communication is conceptualised by three types of communication: intra-project communication; extra-project communication involving organisational liaisons (i.e. transfer of intra-organisational communication by project members); and gatekeepers of information (i.e. transfer of extra-organisational information by external stakeholder-contact personnel) (Lievens and Moenaert, 2000).

Innovation, as in the delivery of projects by project teams, thrives on communication, with development performance being greatly influenced by the quality of the communication process during the development process (Allen, 1985). Both individual team members and the group are the resources that have to be managed towards organisational effectiveness in terms of successfully completing the project. Thus, individual performance will contribute to group performance and this in turn will enhance organisational performance. Consequently, understanding group behaviour as well as individual behaviour is critical for effective project team management. In this context, the group perspective is the project team involving team members drawn from within different organisations.

The importance of internal communication in a service organisation has been widely acknowledged in the literature (Lievens and Moenaert, 2000). Furthermore, the quality of the interactions between the service provider (team) and the client is largely determined by the effectiveness of managing external and internal communication flows during the project development process. Taking this further, effective internal and external communication is seen as a prerequisite for new project success (Thwaites, 1992). Indeed, inadequate internal communication has been identified as a major problem hindering organisations' new service development endeavours (De Brentani, 1989).

Communication and co-operation between the different specialised (functional) project team members during the project delivery process is clearly essential considering the divided responsibility for ultimate project delivery and client satisfaction. Given the definitions of communication presented above, effective project communication may be considered as those changes in information receiver behaviour that were intended by the information source. Such changes may be changes in knowledge, attitude, or in overt behaviour.

In discussing the role of project communication and its subsequent impact, organisations may be thought of as social structures primarily involved in information processing. Adopting an open social systems view assists us in understanding the communication patterns in project teams. The project team consists of specialist persons drawn from discipline-specific organisations, the team members being interdependent as information and work flows are exchanged. The organisational entities to which team members belong are in turn interdependent with some larger environment.

Aside of the project delivery process being considered a process of communication and information processing, it can also be considered to be a process of uncertainty (or risk) reduction (Lievens and Moenaert, 2000). More specifically, the view here is that the greater the uncertainty associated with the project, the greater the amount of information that must be processed among decision-makers to achieve a given level of performance. According to Lievens and Moenaert (2000), acceptance of this principle leads to a contingency approach to managing project communication. As the project team has to process and gather information during the project life-cycle, the project team structure is therefore a critical determinant of the information processing task and innovation activities in particular.

A match between the communication structure and the information requirements is related to higher performance (Tushman, 1979). If the effectiveness of project communication flows is conceptualised by the level of uncertainty reduction, the better the match between project communication and level of work-related uncertainty the more effective communication flows will be in reducing the level of uncertainty. Project uncertainty or risk can emanate from within the project team, or may stem from sources external to the team. Examples of internal uncertainty would be internal conflict, a lack of team member role definition, or poorly defined management structures and communication patterns (Bowen, 1993). External uncertainty includes issues such as a poorly defined client brief, adverse economic conditions, and interface conflict. Internal conflicts are experienced among the project team participants, whereas interface conflicts are typically between the project team and groups outside the project (Awakul and Ogunlana, 2002).

Project team members act as organisational liaisons in mediating communication between their team and the discipline-specific organisations to which they ultimately belong. These project team members effectively act as gatekeepers of information, as they potentially possess crucial market information that should be channelled to their fellow project team members (Bowen and Schneider, 1988). The role of users also needs to be considered. Cairns and Beech (1999) stress the importance of user involvement in project team decision-making, where user feedback is integrated into an interactive cycle of communication and is transposed into 'feedforward' and acting as a positive influence on the design process.

5.5 EXTERNAL STAKEHOLDER COMMUNICATION

External stakeholder communication is an integral part of stakeholder management (see chapter 4). It is not merely a response to the external environment; it also creates the environment. The issue of organisational legitimacy is relevant here. Organisational legitimacy may be framed in terms of the congruence between social values associated with or implied by their activities and norms of acceptable behaviour in the larger social system of which the organisation is a part. Where the two value systems are congruent there is organisational legitimacy. When an actual or potential disparity exists between the two value systems, a threat to organisation legitimacy will exist (Dowling and Pfeffer, 1975). With decreasing legitimacy comes greater pressure to communicate.

External stakeholder participation has become an important component of many decision-making processes for managing environmental and technical risks. Avari, Gregory and McDaniels (2001) argue that values are at the core of all risk issues and should be explicitly accounted for when making decisions about managing risks. Given this, external stakeholders participating in a structured, value-focused risk communication approach will be able to make better informed and higher quality decisions in the context of a risk management problem. Moreover, utilising a value-focused decision structure would make participants more comfortable with their decisions, more satisfied that the selected alternative reflected their key concerns, and more satisfied with their decisions (Avari, Gregory and McDaniels, 2001).

According to Snary (2002), insofar as communication with external stakeholders is concerned, many communication models are dominated by the findings of technical assessments, with only limited opportunity for interested stakeholders to participate fairly and meaningfully in the project development process. In other words, these forms of communication tend to restrict the scope, quality and role of wider technical and non-technical information in the decision-making process. To overcome this problem, Snary (2002) advocates the use of process and outcome criteria.

Process criteria relate to the degree to which (risk) communication enables stakeholders to participate fairly and competently; outcome criteria relate to the extent to which risk communication has substantive benefits (for the project planning process). Process criteria may be thought to include issues such as: the representativeness of the participants; the effectiveness of the contact group model; and the compatibility of the process with the objectives of the participants. Outcome criteria include: the degree of knowledge among the participants; the impact of the participants; the promotion of trust; and the resolution of conflict.

Mearns, Flin and O'Connor (2001) report the increasing importance of the role of context and culture in shaping both the perception and experience of risk, leading to different 'worlds of risk' in organisations (risk subcultures). If risk is not a unified phenomenon, how do we achieve shared attitudes, perceptions, and beliefs with respect to risk? Moreover, in the presence of these differing 'worlds of risk', how do we effectively communicate risk issues?

5.6 RISK PERCEPTIONS AND THE COMMUNICATION OF RISK

We have seen that information, taken in its broader sense and pertaining more to knowledge than to data per se, can only be transferred between parties if that information is transmitted (and received) in a manner which is meaningful to both. Clearly, the more closely the models of reality of the transmitter and the receiver correspond, the more effective becomes the resultant communication. To reiterate, the effectiveness of the project team partnership is a function of the effectiveness of the communication between the internal stakeholders. In building procurement projects, for example, communication constitutes a major problem (Griffith, 1985; Bowen, 1993). This problem is compounded by the increasing partition of the building procurement process into more and more specialised disciplines, resulting in little mutual appreciation of each other's knowledge base. The purpose of communication between project stakeholders, and the context within which that communication occurs, are as important as the communicators themselves.

Risk perception

A little-acknowledged aspect of the risk communication context is that, for every project stakeholder, perception (i.e. knowledge and understanding) of risk and risk management is rarely acquired through first-hand experience. We rely on a variety of media – and thus the communicating power of others – for our information; e.g. lectures, videos, audio tapes, case studies, journals, conferences, and textbooks. We learn about risk largely at second or third hand; and then use this learning to inform our first-hand communication and decision-making about risk in the real-life project environment. If the risk learning messages distort risks, e.g. exaggerating some risks or diminishing others, or use inconsistent descriptive framing for risks, then our subsequent decision-making judgment is likely to be at least clouded, or even skewed.

The social perspective of risk (see chapter 2) cautions against an overreliance on exclusively mathematical approaches to assessing and modelling risk, and against overly mechanistic 'hard systems' methods of managing risk. It also places a greater emphasis on risk communication, and brings into focus the role of professional judgment in assessing and managing risk. Faced with a diversity of views about risk, we seek the advice of experts we can trust. We assume that they will understand, and be capable of dealing with, perceptual dissonances. However, we must recognise that experts too are susceptible to perceptual heuristics and biases (Mak, 1992; Birnie,

1993). We also assume that they are capable of adequately communicating (to the appropriate receiver(s), utilising the appropriate channel and medium of transmission, and conveying the desired message content), without imposing their own risk value system. This may be an overly optimistic expectation.

The communication of risk

In the early stages of risk communication research and practice, factual knowledge of risks was believed to be a primary predictor of support or opposition, but more recent research has shown this not to be the case (Heath and Abel, 1996). For example, members of the public affected by risk do not readily accept the point of view of companies or governmental agencies on the degree of risk. Heath, Seshadri and Lee (1998), in studying the communication of risk associated with communities living near hazardous chemical plants, found that knowledge about risky situations correlated positively with trust, perceived openness, and support from the chemical company, and negatively with cognitive involvement, uncertainty and dread. Heath, Seshadri and Lee (1998) found that variables in risk communication comprise: trust, involvement, uncertainty, openness/accessibility, knowledge, and support/opposition.

Otway (1992, cited by Palanchar and Heath, 2002: 130), comment on the complexity of risk communication, stating:

> **Risk communication requirements are a political response to popular demands. The main product of risk communication is not information, but the quality of the social relationship it supports. Risk communication is not an end to itself; it is an enabling agent to facilitate the continual evolution of relationships.**

The general principles associated with risk communication apply no less within the context of risk communication within and beyond the project team.

Thomson and Bloom (2000) provide useful guidelines for the communication of risk. They argue that, to better inform project (risk) managers, risk assessors must appreciate and present the broader context of the decision, and must convey how uncertainties and weaknesses in the assessment may influence stakeholder perceptions of risk and the effectiveness of different risk management options. Key uncertainties (dominant contributors to risk) should be highlighted, as should data gaps, to allow the risk manager to assess the level of confidence in the assessment. Moreover, those uncertainties or risk areas that might be the focus of external stakeholder attention and possible reaction need to be identified and discussed, together with possible remedial action.

In summary, effective communication about risk is of paramount importance. Effective risk communication requires not only placing appropriate emphasis on the risk messages themselves, but also giving attention to the message media, the message senders and receivers, and the provision of feedback to ensure that mutual understanding of project risks has been achieved.

5.7 CHAPTER SUMMARY

In this chapter, we have discussed issues of decision-making, risk perceptions, interpersonal communication, and the communication of risk messages, all within the general context of organisations and project teams.

It is clear that, within the complex dynamics of decision-making in organisations and project teams, the risk perceptions of project participants are different and that communication is an integral part of decision-making. Consequently, the effectiveness of the organisation and any project team is a function of the effectiveness of the communication process within the organisation or project team. Given this, appropriate attention needs to be focused on the fidelity of risk communication messages. More specifically, risk communicators need to be aware of the importance of the message senders and receivers, the medium of transmission, the risk message content, and the necessity for feedback to establish the degree of shared meaning if mutual understanding and appreciation of project risks is to be achieved.

In the following chapter the processes of systematic risk management are introduced, providing the linkage between the theory and practice of risk management.

CHAPTER 6

SYSTEMATIC RISK MANAGEMENT

6.1 INTRODUCTION

For the great majority of project stakeholders, risk is too important to their continuing survival and success to be ignored or dealt with haphazardly. Modern society's expectations of corporate behaviour and public accountability demand that organisations give due regard to the risks they face (or which they create for others). Accountability has become a critical facet of the ways in which organisations operate in the modern world.

Given this trend, and the increasingly complex and risky nature of many projects as discussed in earlier chapters, recognition of the need to adopt and maintain a systematic approach to project risk management is inevitable, and permeates public and private sectors alike.

This chapter is the fulcrum of the book, providing the essential turning point between theory and practice. It sets out the processes of systematic risk management, commencing with arguments for the proposition that project risks are more correctly perceived as the risks of the stakeholders involved in a project. Justification for a systematic approach to risk management is given, along with an explanation of its benefits and implications. The early stages and procedures for a typical risk management system are then described in greater detail.

Our aim in this chapter is to show how the material dealt with earlier – risk, projects, organisations, communication and decision-making – are integrated into an appropriate system for dealing with risk.

6.2 STAKEHOLDER RISKS – NOT PROJECT RISKS

It is convenient to refer to 'project risks' in any discussion or exploration of the myriad of things that can happen with projects. Indeed, we have made such references throughout this book (and do not apologise for them!). The brevity and familiarity of the term outweigh any lack of validity it might have, but it is important that terminological convenience should not take place at the expense of achieving a proper understanding of the ownership of risks and the responsibility for managing them.

In chapter 2, the point was stressed that risk is a social construct, experienced and understood by people. Projects, by definition, are impersonal undertakings and thus cannot experience risk. Projects do not make decisions about the tasks, technologies, resources or organisation required for them. People make those decisions. The risks associated with projects are therefore more properly those of the people (organisations) involved with those projects. Nowadays it is customary to identify such people or organisations as project stakeholders.

Even if a project stakeholder is a large corporate entity, it still has a persona – indeed legally it will be treated as such – and is thus capable of experiencing risk. Furthermore, since risk is associated with project decision-making, it is the decision-makers within the stakeholder organisation who will experience risk most directly, and who should be closely involved in managing it.

Conceptually, therefore, it is incorrect to refer to 'project risks', and 'risky projects', unless in the context of a specific project stakeholder. What is a project risk for one stakeholder may not be so for another, and a project which one stakeholder regards as risky may be completely straightforward and innocuous to others.

Some projects are undertaken wholly within an organisation. This 'in-house' perspective is sometimes known as 'enterprise project management' (EPM). In EPM, projects may be conceived, resourced and carried out by a single organisation utilising its own project management expertise. Even with EPM however, it is rare for every project to be completed wholly within the boundaries of the organisation. External suppliers, contractors and subcontractors are frequently involved to some degree. In accordance with the proposition we have argued, each external participant then becomes a stakeholder, responsible for managing the risks to which it is exposed on the project.

Figures 6.1 and 6.2 illustrate this proposition diagrammatically. Figure 6.1 suggests that risk management is one of a number of organisational management processes that each stakeholder may apply to its activities and thus to each project with which it is involved.

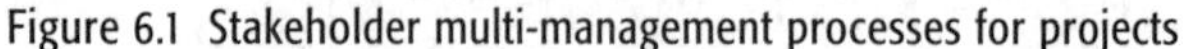

Figure 6.1 Stakeholder multi-management processes for projects

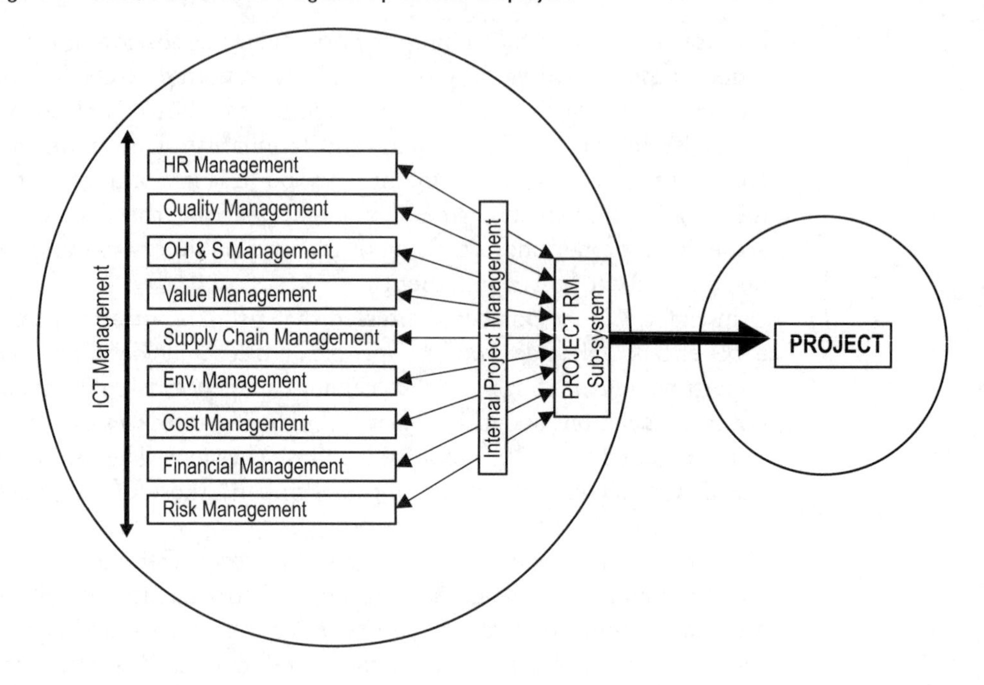

STAKEHOLDER ORGANISATION

Emanating from the multi-management processes of figure 6.1, each stakeholder should devise risk management sub-systems to bear upon the projects in which it is involved. While a sub-system may be project specific, it should clearly lie within the framework of the broader risk management system of the stakeholder. This proposition is given fuller consideration in chapter 7.

In figure 6.2, the multi-management perspective has been widened to reflect this situation. A project may include multiple stakeholders, each needing to apply its own risk management processes. Similarly, at least some of those stakeholders will be involved in multiple projects at the same time. Their organisational risk management structures must obviously encompass this situation. The nature and intricacies of this concept of multi-management processes are really beyond the scope of this book, and are therefore not discussed further here.

Not all stakeholders maintain a consistent participation throughout every phase on every project in which they are involved. According to the nature and demands of a project, some will discontinue their involvement at various times during any of its procurement, operational or disposal environment phases. Other stakeholders might replace them. During the procurement phase of a construction project, for example, a specialist piling contractor commences on-site work early in the construction process, perhaps preceded by an excavation contractor whose task is to clear and level the site. When the piling work is completed, the specialist contractor leaves the site (usually for good) and further excavation commences – but this time for the more intricate digging of pits and trenches for footings and services. Formwork constructors, steelwork fixers, concretors, plumbers and others follow, according to a pre-planned methodological sequence of construction activities. Eventually the building begins to rise out of the ground, gradually taking on more recognisable form and appearance. These 'ebbs and flows' of the 'actors' in the lives of projects are an important reminder of their dynamic nature. A stakeholder risk management system should be equally dynamic. Indeed, there can be no such thing as a static system of risk management – it would constitute a contradiction in terms of effective management.

Figure 6.2 Multi-stakeholder projects, multi-project stakeholders, and risk management systems

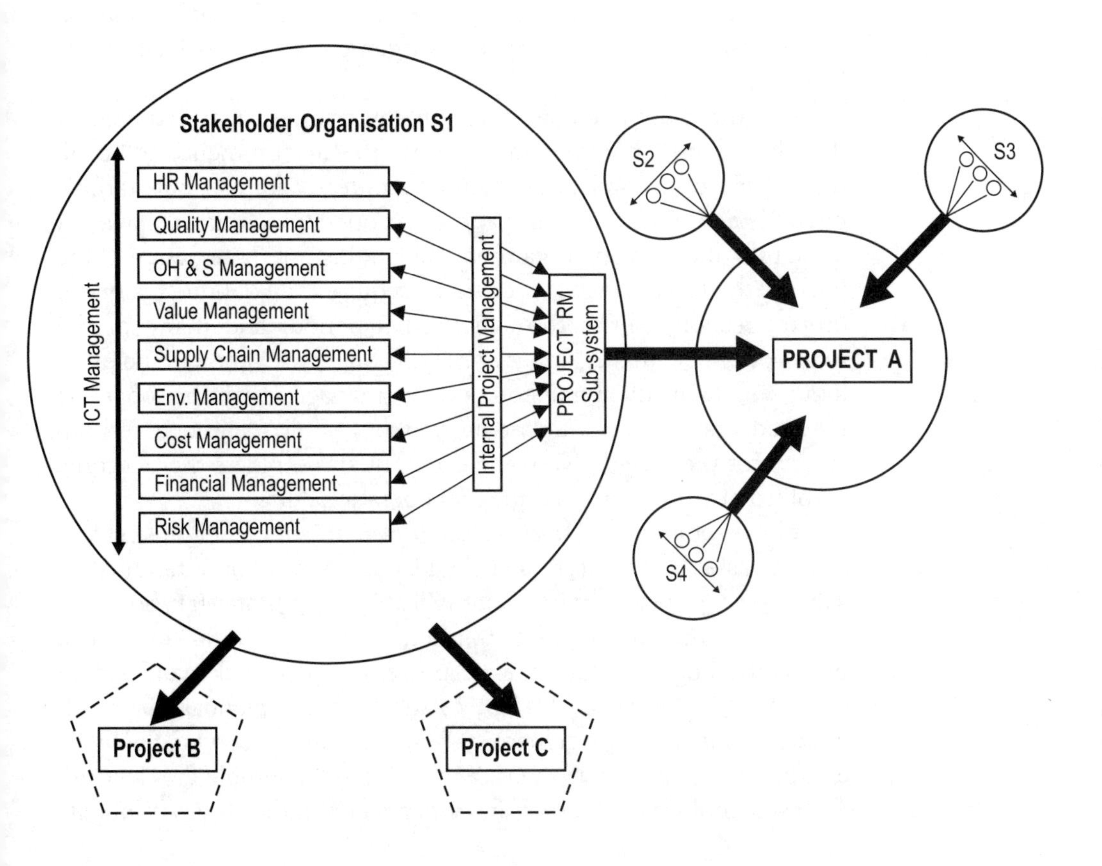

Another important point to note in all of this is the role of project managers. Figure 6.1 indicates a level of project management occurring within any project stakeholder organisation, and this is usually the case, whether or not the person taking a leading role in this capacity is actually called a project manager. For many projects, however, a separate person or organisation is appointed to manage the project on behalf of the key stakeholder (owner, client, project proponent), often on a fee-for-service basis. It is important to clarify risk management responsibilities in such situations. An independent consultant who is providing project management services is clearly exposed to risks in his or her professional capacity, and should therefore take steps to deal with them (in order to protect his or her own interests). Inevitably, however, the relationship between client and project manager means that the latter will also become involved in the client's risk management processes. Indeed, the project manager may even be called upon to instigate a formal risk management system for the client. Any such arrangements should be clearly explained in the service agreement, so that the rights of each party, and the individual integrity of their risk management systems, can be maintained.

This issue rarely arises in projects where the project management function is delivered through EPM, since there is no legal separation of the parties and only one risk management system is likely to be involved.

Beyond the procurement environment, a project may require further design input before its operational life can commence. A hospital project, for example, may be built and ready for use, but staffing, movable equipment, supplies and other operational factors have yet to be fully planned and organised. Depending upon the nature of the project, the project manager could be required to continue his or her project responsibilities throughout this period, and many of the issues of project management risks will still be relevant. On the other hand, the distinctive inputs of a project manager might no longer be required, and the client immediately assumes full operational responsibility for the facility. New and different stakeholders then become involved at various times during its operational life.

At some point in its life-cycle, the project might come to be regarded as an asset (or liability!) to be liquidated, disposed of or terminated. This project disposal environment will attract yet more stakeholders.

The argument behind all of this is that, throughout its life-cycle, a project itself does not (and cannot) possess a unique risk management system. There is no equivalent of the 'owner's manual' for a car, whereby one risk management system is capable of serving the interests of everyone involved with a project over its whole life. Instead, each stakeholder – whatever and whenever and for however long its involvement and participation in a project – must choose how to manage the risks arising from that involvement.

This is the reality of all projects. For the purposes of this book, however, and in the interests of brevity and simplifying understanding, discussion about project risks and their management will be regarded as proceeding in the context of a hypothetical stakeholder whose interests lie in a particular environment of a similarly hypothetical project. This is actually part of the topic of system boundaries that we will necessarily have to return to later. First, however, it is important to examine the essential features of a risk management system.

6.3 FEATURES OF SYSTEMATIC RISK MANAGEMENT

We all manage risk. As individuals we tend to do this intuitively most of the time, rarely exercising a deliberately cognitive approach. While intuition might suffice for an individual person (and we explored in chapter 2 just how difficult that might be, even for deciding to cross a road), it is hardly likely to be adequate for an organisation seeking to achieve specific objectives in a project. As we have noted already, projects are becoming increasingly complex; the rate of technological change has become more rapid; expectations of accountability have become more demanding. It is more and more necessary to cope with volatility and turbulence in almost every aspect of organisational activity. Status and reputation have to be protected. National and international incidents have lead to greater recognition of local and global vulnerability. All this has contributed to the development of a broader conceptual view of risk among communities of all kinds, and to recognition that formal procedures are required to deal with it.

We could adopt an ad hoc attitude to risks, examining and treating each one as we become conscious of its existence and potential impact. While this might be acceptable on a personal level, from an organisational perspective it is unlikely to be sustainable in the long term.

A systematic approach to risk management increases the capacity of an organisation to handle risks at all levels. It promotes internal transparency and common understanding of the business activities of the organisation, and facilitates the establishment and growth of an organisational culture of, and a commitment to, managing risks. Having a formal risk management system in place simplifies the intra- and inter-organisational capture and transfer of risk knowledge. It provides the means for assessing best practice and for benchmarking management performance, and establishes a platform that permits the benefits of modern information and communication technology to be exploited.

An effective risk management system delivers a number of benefits. Within a stakeholder organisation, congruence between the organisation's objectives and the project objectives can be confirmed, or any conflict between them identified. Trust and confidence between the levels of management, and between management and operatives, is improved. Responsibilities are clarified, and ownership of responsibility more willingly accepted. Potential problems are exposed to a wider internal audience. The organisation's capacity to deal with new risks is enhanced. More focused risk training is possible. Creative approaches to handling risks can be encouraged and developed. The nature and scale of potential crises is better understood, allowing better visualisation of more distant time horizons. Anticipation replaces surprise, and informed thinking takes over from reliance on luck. Above all, a systematic approach to risk management should encourage decision-making within an organisation that is more consistent, more controlled and yet at the same time more flexible.

6.4 RISK MANAGEMENT SYSTEMS

A good risk management system (RMS) will allow an organisation to look forward to the future for each project, while maintaining a convenient capacity for looking backwards to the wisdom gathered from its previous project risk experiences. This might entail the adoption of a dual-system approach. On the one hand would be dynamic, project-specific systems as the vehicles for managing the risks of each project. On the other hand, there should also be an organisational meta-system, containing a risk register structured as a repository of organisational knowledge of project risks.

There should also be a third system in place. This would have an organisational – as distinct from a project – macro-focus, in order to encompass the risks associated with the internal tasks, technologies, and resources necessary to sustain the organisation itself as an ongoing concern, i.e. to deal with the likelihood of events occurring which could adversely affect the continuing life of the organisation. As noted earlier however, we will concentrate here upon project risks and project RMSs.

At a project level, the RMS will comprise the means for identifying competing interests, and employ techniques for weighing up inadequate information about the project. It will take note of risks already allocated or risk treatments already in place. It will provide for the recording of decisions for action, and the assignment of responsibilities. The systematic nature of the RMS will be inherent in the definitive processes it incorporates, the approaches to data analysis it employs, the treatment regimes it embraces, and the usefulness of the information and knowledge that it captures.

Stages in risk management

The literature describes numerous process stages as essential to an effective RMS. Despite minor disagreements on terminology, and occasional combinations or separations of some stages, there is a large degree of agreement between nearly all authors about them. It is safe to say, therefore, that in principle a good RMS for a project should comprise processes to:

- establish the appropriate context(s),
- identify the project risks the stakeholder organisation will face,
- analyse the identified risks,
- develop responses to those risks,
- monitor and control the risks during the project, and
- permit post-project capture of risk knowledge.

These processes can be represented as flow process diagrams, as shown in figure 6.3. The RMS is deliberately shown here as a cyclical loop, to indicate a learning process that should be ongoing from one project to another.

Figure 6.3 A systematic cycle of risk management

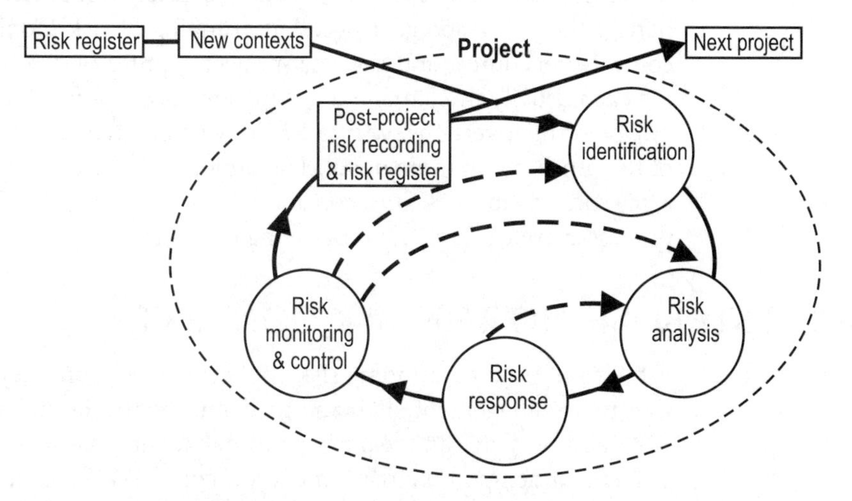

As an organisation engages in a new project, it accesses its repository system of risk knowledge: the generic project risk register. Possible structures for this will be discussed later in this chapter. The risk contexts for the project are established. A process of risk identification is undertaken, and the identified risks are then analysed in greater detail. Once risks have been identified and analysed, alternative risk treatments can be explored and decisions made about the preferred response for each risk. Procedures for monitoring and controlling risks during the ongoing progress of the project are put in place. Return loops to earlier RMS processes may be necessary during the monitoring phase, as new risks are identified or as the circumstances of known risks change. Eventually, the risk knowledge gained on the project is captured, transformed into a suitable format and recorded in the organisational risk register. The whole cycle is then ready to begin again on the next project.

Within this cycle, risk identification generally takes place as a single, congruent process – at least in the initial part of the cycle – to identify as many risks as possible. Risk analysis and risk response processes however, could also be undertaken separately, or jointly for each risk in turn. This is because the exploration of alternative responses for a risk could require additional analytical inputs. There are no firm rules or guidelines as to whether or not to keep the risk analysis and risk response stages entirely separate. One point to bear in mind is that the risk response stage finishes with clear decisions about risk treatments, and therefore requires the participation of the appropriate decision-makers in the organisation. On the other hand, the analysis of identified risks in greater detail is not necessarily the task of decision-makers. Indeed, it is sometimes necessary to call on the help of risk analysis experts from outside the organisation. In practice, therefore, risk analysis is usually done separately from at least the final part of the risk response stage. The planning of risk monitoring and control procedures, and the post-project capture of risk knowledge, are each usually implemented as separate stages of risk management.

Given this overall view of a RMS, we can begin to examine each of the processes in more detail. The project risk context and risk identification techniques are considered in this chapter. The other processes are dealt with in following chapters.

6.5 ESTABLISHING THE RISK CONTEXT

Establishing the appropriate risk context is one of the keys to effective risk management. It helps to delineate the boundaries of the RMS and provides an ideal trigger to risk identification.

In deciding upon the risk context (or contexts), a balance has to be struck between contexts that are too wide, and those that are too narrowly focused. Contexts framed as 'ICT project to update bank ATMs' or as 'New shopping centre in the High Street', for example, are almost certainly too wide. 'Tightening bolts with torque wrench', on the other hand, is likely to be too narrow.

Objectives

Clues for choosing risk contexts have already been provided in the earlier chapters of this book. One approach might be to start with the organisation's objectives for the project. Where multiple objectives are sought, each of these should be examined in terms of the more detailed plans established to achieve it. These will provide the basis for identifying risks that could threaten that achievement. Also intrinsic to the notion of multiple objectives is a hierarchy of levels for objectives. This is illustrated in figure 6.4.

Figure 6.4 A hierarchy of project objectives

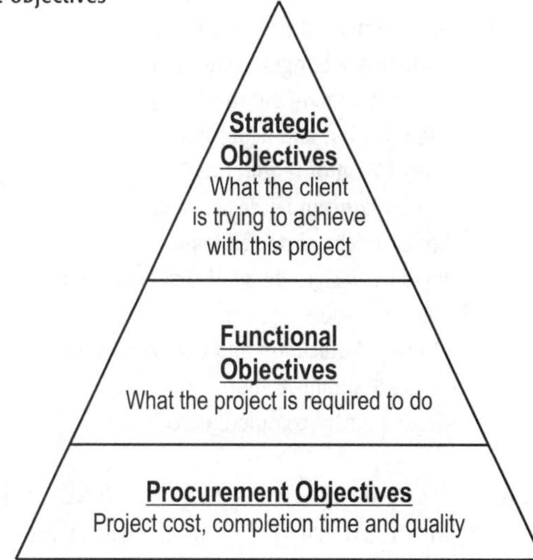

From the perspective of a client stakeholder, most objectives at a procurement level relate to time, cost and quality: the time required to get the project delivered ready for use; the cost of getting it ready for use; and the quality needed to make it fit for use. At a functional, or operational level, the client will have expectations about what the project should do – an ATM project for a bank might be expected to deliver a 24-hour cash dispensing service to customers with 95 per cent reliability, 99.95 per cent accuracy and 98 per cent security. Beyond this level there may be strategic objectives sought for the project: what the stakeholder wants to achieve with the project. For the banking client this might be gaining market dominance in ATM penetration in a particular region. Not every stakeholder organisation will have the same objectives, nor even the same number of hierarchical levels of objectives, as other stakeholders in the projects in which they are involved. This reinforces the proposition that risk management systems are stakeholder, rather than project, specific.

Project environment

The appropriate project environment provides another contextual starting point. In terms of the project life cycle, is the stakeholder organisation involved in the procurement environment, the operational or disposal environments, or some combination of these? Each relevant environment is then examined and broken down into the process stages or plans necessary to deliver it. Thus, for the procurement environment for a proposed new hospital, these process stages might include:

- confirmation of functional needs
- confirmation of funding capacity
- appointment of project manager
- preliminary budget establishment
- outline planning approvals
- site selection and acquisition
- preparation of requirements brief
- call for conceptual design proposals
- appointment of design team
- building design, detailed cost plans and tender documentation
- call for tenders
- tender adjudication and contract award
- contract administration
- occupation of completed facility.

Note that this assumes that the RMS is being implemented for the hospital client organisation, and that the building procurement system adopted is of the traditional separated design-tender-construction type. For other stakeholders, or with alternative procurement systems, these stages would be different. Each stage then becomes an identifiable context for risk management. Note too that the stages are not strictly sequential in a temporal sense: several could overlap, so that combinations of stages might form more appropriate risk contexts.

Project elements and sub-elements

As an alternative approach to establishing the stages of each project environment, the breakdown of any of the appropriate environments into its task, technology, resource and project organisational elements and sub-elements, could provide a suitable contextual basis for risk management. Basically, this approach would follow the propositions set out in chapter 3. Careful control is needed with this approach, lest it is allowed to descend into a morass of detail and the RMS becomes unmanageable.

What might have become evident in all of this is that, in principle, establishing the contexts for risk management is largely a matter of tracking the decision points for a project. Logic confirms this, since we have already stated that it is project decision-making which gives rise to risk.

When the risk context has been established, it is possible to proceed with the identification of particular risks.

6.6 RISK IDENTIFICATION

A satisfactory process of risk identification is crucial to effective risk management, since unidentified risks cannot be systematically managed. Yet they remain risks. It is therefore worthwhile for any organisation interested in implementing a RMS to spend time in considering how best to undertake this process.

Essentially, risk identification sets out to answer the question: what could threaten the satisfactory achievement of this objective, the completion of this task, the application of this technology, the acquisition of this resource, or the performance of this organisation? Or, to put it another way: what could happen to make this project decision a bad one?

Note that the emphasis here should be on the risk event, rather than its consequence: on cause rather than effect. This is not to say that impact plays no part in risk identification. Many people intuitively use it as a starting point in the process. Cost and time overruns, for example, are often suggested initially as risks to projects, but a little thought will show that these are actually consequences of prior risk events and the causal factors or situations must then be explored.

RMS boundaries

The event/consequence dilemma in risk management leads to the issue of system boundaries in project RMS. Establishing system boundaries helps in distinguishing risks that are *exogenous*, i.e. are derived from outside the system, from those that are *endogenous*, i.e. occurring within the system itself. Both types will impact upon the project, but the organisation's capacity to manage them will be affected.

Establishing appropriate contexts and determining the boundaries of its project risk management system are important matters for each stakeholder to consider. This consideration, however, should be made well before the stakeholder becomes involved in a project. Ideally, it will have been part of the organisation's processes in building a risk management system, as we shall see in chapter 9.

Given that a project stakeholder has established the risk management contexts, and is at least aware of the system boundaries for its RMS, how should it proceed to identify the risks to which it may be exposed?

Identification approach

All that we have said earlier in this chapter, about ownership of the RMS, should not be seen as precluding a team approach among project stakeholders to the task of identifying (and assessing) project risks. Eventually, of course, each risk must become the responsibility of the stakeholder to whom it has been allocated, but a project team workshop can be a useful approach to identifying and allocating risks.

In regards to a recent Australian project involving the extensive redevelopment of an existing mainline city railway station, for example, a workshop approach was adopted to explore the risks associated with one particular aspect of the project. The risks issues concerned the reconstruction and extension of the station roof over the existing platforms. The main contractors for the project were required to undertake most of the heavy erection work for the new roof trusses at night, allowing the station to continue to function during daylight hours. Space constrictions, and the size and mass of the roof components, prevented the use of conventional mobile cranes for the erection process. It was proposed that flatbed railway wagons should be adapted to accommodate a special heavy-lift crane, spanning across three sets of rail tracks, and winched backwards and forwards along the tracks over the required area. A one-day workshop, facilitated by risk management specialists, was held to explore the practicality of this idea and to identify the associated risks. Stakeholders represented in the workshop included the main contractor, project manager, railway operating authority, railway engineering authority, roofing subcontractor, crane specialist, structural engineering consultant, and architectural consultant.

Clearly, a single workshop such as this could not be expected to deal with more than a limited number of issues (the roof structure erection process). Clearly too, the workshop was limited to the identification and assessment of the risks associated with this hoisting tasks of this process. Further stages of the risk management process would remain the responsibility of the relevant stakeholder in each case.

Within a workshop approach, a variety of risk identification techniques may be used.

Identification techniques

Several risk identification techniques exist. The ingredient common to virtually all of them is brainstorming. Currently, no automated techniques of risk identification exist, at least for projects. At relevant intervals in the risk identification process, the people involved should be asked to respond to the question we raised earlier: what could happen in this particular context to threaten a satisfactory outcome for this organisation on this project? The process is therefore iterative to some extent.

Risk identification brainstorming could be used on its own in a completely unstructured manner, but this would hardly accord with an organisation's desire to be systematic in its risk management. In fact brainstorming is easier, and usually more successful, when it is guided in some way, i.e. applied within more structured identification techniques. We can now look at several of these.

CHECKLISTS

Checklists are often advocated as a technique to stimulate brainstorming in risk identification. Where the lists are derived from a stakeholder's organisational project risk register (i.e. a compendium of risk knowledge gleaned from previous projects), there is some merit in them. However, they are rarely entirely sufficient for effective risk management since they are unlikely to help in identifying new risks that have not previously been experienced. Lists also require frequent updating in terms of relevance and currency, particularly where an organisation's business processes have substantially changed over time.

Many of the lists prepared for other project purposes are less adequate for identifying risks. Examples might include people lists; information needs lists; and lists of project milestones. Of these, only the first has direct application as it relates to the organisation element of projects, but it needs to be placed in a more coherent contextual framework.

As a somewhat crude analogy, shopping lists tell us little about the risks of shopping, since they usually only indicate what we want to buy, and rarely explain where to buy, when to buy, how to buy, how much to buy, what not to buy, or how to get everything home afterwards! Lists therefore have limited usefulness in risk identification, unless they provide sufficient focus.

RISK SOURCE CATEGORIES

Risk source categorisation was suggested in chapter 2 (see figure 2.3) as a way of classifying risks. This alternative form of list can provide a more structured basis for brainstorming questions along the lines:

what risks of this type could arise on this project? However, while this ensures that each type of risk is considered on each project, it does not mean that all risks within a particular risk category will be identified.

DECISION POINTS

Since we noted in chapter 3 that risk arises from project decision-making, tracking project decision points should simplify the process of risk identification. To an extent this is true, but only within a more highly organised framework of where, when, how, and about what the decisions are made.

PROJECT ENVIRONMENTS AND ELEMENTS

Project environments on their own are insufficient for identifying project risks as their perspective is usually too broad, particularly where a stakeholder is involved in only one of the procurement, operational or disposal environments of a project.

Within each project environment, the task, technology, resource and organisation elements, and more particularly their respective sub-elements, should provide an effective focus for risk identification. These too may be in list format. The sub-elements within each element are important for risk identification as they are capable of exposing the levels of differentiation and interdependency associated with them. Furthermore, they point to the loci of decision-making and the inherent uncertainties attached to many project decisions. Remember that decision-making and uncertainty lie at the heart of risk.

Sub-element/risk source category matrices

Given the arguments presented so far, the two-dimensional project sub-element/risk source category matrix mapping technique described in chapter 3 (table 3.2) is probably one of the more effective approaches to identifying project risks, especially if combined with an organisational project risk register. For each sub-element in the left-hand column of the matrix, participants in the identification process are asked to consider each risk source category and suggest (brainstorm) what risk events under that category might be associated with that particular sub-element. Care must be taken with this technique. The sub-element analysis must not become so extensive that participants are faced with too much detail and lose confidence in ever finishing the risk identification process.

OTHER MATRIX APPROACHES

The element/sub-element project breakdown lends itself to other matrix approaches for identifying risks. The row breakdown of elements and sub-elements is retained but, instead of risk source categories as the horizontal column variables in the matrix (table 3.2), these could be replaced by variables representing each of the people or departments associated with the project within the organisation. Cell entries would then reflect which parts of the organisation are involved with each element and sub-element; and what could threaten that involvement. Multiple entries across each row would indicate the extent of intra-organisational communication necessary to deal with those risks.

In a similar manner, the horizontal column variables could be used to reflect the different external stakeholders in the project. This matrix would thus reflect identification of inter-stakeholder risks.

The attractiveness of matrix forms of risk identification is that they are easy to prepare and lend themselves to the use of structured brainstorming sessions for completing them.

HAZOPS

HAZOPS (hazard and operability studies) is a risk identification technique originally developed for the process industries such as the petro-chemical industry. As such, it is particularly useful for projects involving the delivery or operation of production and processing facilities. However, this does not restrict the use of HAZOPS just to chemical engineering projects. Remember that hospitals also adopt flow process approaches in designing their operational modes.

HAZOPS uses an interrogatory form of risk identification. Experts in the relevant field (usually constituted as a panel) interrogate every part of the process design for the whole facility. Their questions are framed according to the nature of the facility, the purpose of the components being examined, and their expert knowledge of the processes involved. Table 6.1 shows typical questions for a section of a hypothetical processing plant project.

While the HAZOPS technique can be adapted to provide a basis for risk identification in other types of projects, it is best suited to flow processes that are relatively straightforward and linear in nature. It identifies risks that are for the most part technical, and is therefore less suited to situations that involve other types of risks. Operator error, for example, is often not easily detectable through HAZOPS. Stakeholders in complex projects with high levels of differentiation, and especially with high interdependencies and uncertainties, would probably find the technique too time-consuming and expensive to apply. The task of framing the questions would itself be a considerable undertaking.

HAZOPS also assumes that all (or almost all) of the detailed project design is already complete. This makes it unsuitable for risk identification in the early stages of many projects.

Table 6.1 Typical HAZOPS approach to risk identification

Item	Reference	Specification	Question	Guide	Situation	Result
1	Dwg. A/21/123	Section: A21	What if...	no	flow occurs through flask AF1?	Auto-sensor AS12 will close Valve AV9.
2	Dwg. B/78/456	Section: B05	What if...	more	pressure than specified occurs at Valve BV6?	Pressure compensator will adjust flow and alert operator panel B/OP/9.
3	Dwg. C/51/789	Section: C48	What if...	less	temperature than specified is generated at Purifier CP2?	Panel alert needed (Design Mod C/51/CP2A).
4	Dwg. D/032/099	Section: D03	What if...	reverse	flow occurs at Filter DFL12?	Non-return valve at DTA16 will protect tank.
5	Dwg. E/83/004	Section:	What if...	as well as	XX waste, YY overspill chemical also flows in channel EC9?	XX is inert. Separator chamber at ES12 will intercept YY.
6	...	...	What if...	...	...	...
7	...	...	What if...	...	...	...

It is appropriate to reflect that the HAZOPS technique could be used in other process-based areas of stakeholder management, such as quality assurance (e.g. design checking) and health and safety management.

FMECA

Like HAZOPS, the FMECA (failure mode and effects criticality analysis) technique was developed for a particular industry, in this case manufacturing engineering. Companies in the automotive industry adopt this approach largely because it is possible to apply FMECA uniformly across a large number of subcontract component manufacturers and suppliers, and require them to respond to parts failures in a standardised manner.

FMECA also needs inputs from experts, particularly in establishing the criteria for faults and failure and setting component manufacturing tolerance standards. In applying the technique, it is usual to list all the components in the relevant system, describing their required function. All possible failure modes and their causes are then explored, particularly with respect to the effects upon the rest of the system and the severity or criticality of these consequences. Alternative corrective measures are considered.

A typical page format for a FMECA risk identification process is shown in table 6.2. The application of this technique actually extends beyond initial fault risk identification through analysis and into the treatment plan, thus effectively creating a mini-RMS for every component of the final manufactured product. The FMECA technique is also used for quality assurance.

As with HAZOPS, FMECA requires a complete product design to be in place in order to be applied effectively, and so suffers the same disadvantages as HAZOPS as a generic risk identification tool for project-

Table 6.2 Format for FMECA approach to risk identification

Item	Component reference	Revision no.	Identity	Mfg. tolerance	QC sampling	Potential fault mode	Failure probability	Failure effect	Treatment plan	Responsibility
1	GT2046789T	MV3097	LF subframe assembly	Specif. SFLF55077	1:1000	Tube spot weld failure	1: 50 000 (crash, collision?)	Serious fillet weld XR insp. (1:500)	Full butt or Prod. eng. QM Costing	Des. eng.
2										
3										
4										
5										
6										
7										

based risk management. The emphasis on technical systems means that the risk of human error may be overlooked. While differentiation complexity is dealt with, interdependency is less adequately treated.

FAULT TREE ANALYSIS

Fault tree analysis (FTA) may be used as a risk identification technique or as a risk analysis tool. It is used to deductively explore underlying causes for an event. When these rational causes have been identified, the top event actually becomes a consequence of one or more of the lower events. In project risk management, this approach often has the benefit of clearly demarcating the project boundaries of the stakeholder organisation's RMS. The example shown in figure 6.5 indicates a top event as the risk of automatic loading doors failing to close properly for a vehicle ferry. Logic is used to trace potential causes for such an incident. The dilemma is where to stop the tracking process. Should 'blocked fuel lines' be attributed to the ingress of foreign objects into the fuel system? Is the faulty wiring due to rodents eating the insulation from cables? Are damaged or missing components the result of deliberate attack or theft? This is where setting the RMS boundary is important, but this can only be done by each unique stakeholder organisation in a project.

If the FTA risk identification approach in the example of figure 6.5 is being undertaken on behalf of a shipyard subcontractor responsible for the loading door installation on a project involving the construction of a new ferry vessel, then the fault tree as shown probably lies wholly within that organisation's project boundaries. On the other hand, if risk identification is being done by a shipping company planning the introduction of a new ferry service, then the top event is probably the risk trigger event in the context of its operational procedures for the ferry. The shipping company, as a key stakeholder in the ferry service project, is vitally concerned with the consequences of this risk event (rather than its causes) and will place it at the boundary of the organisation's project RMS.

In using FTA for risk analysis, probability expressions might be determined for each of the nodes on the diagram, to represent the chance attributable to any of these events occurring.

EVENT TREES

Event tree analysis (ETA) adopts the opposite approach to FTA. Inductive logic is used to explore the consequences of a trigger event. Figure 6.6 shows how this might occur for the shipping company in the ferry loading door example.

Note how this approach explores the consequences of the risk event in an escalating manner, with outcome exits at each level.

Figure 6.5 Fault tree risk identification technique

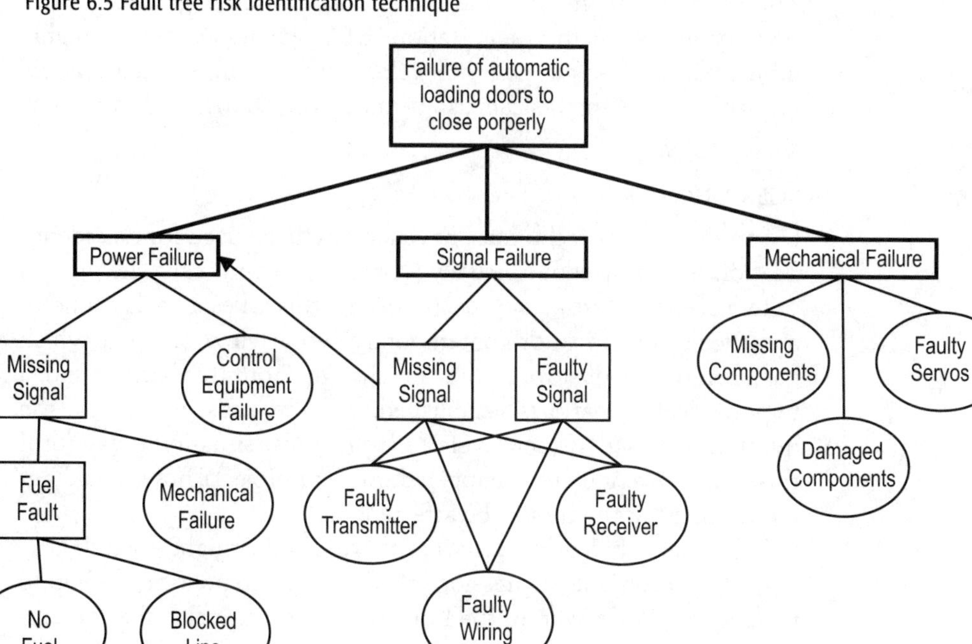

Figure 6.6 Event tree risk identification technique

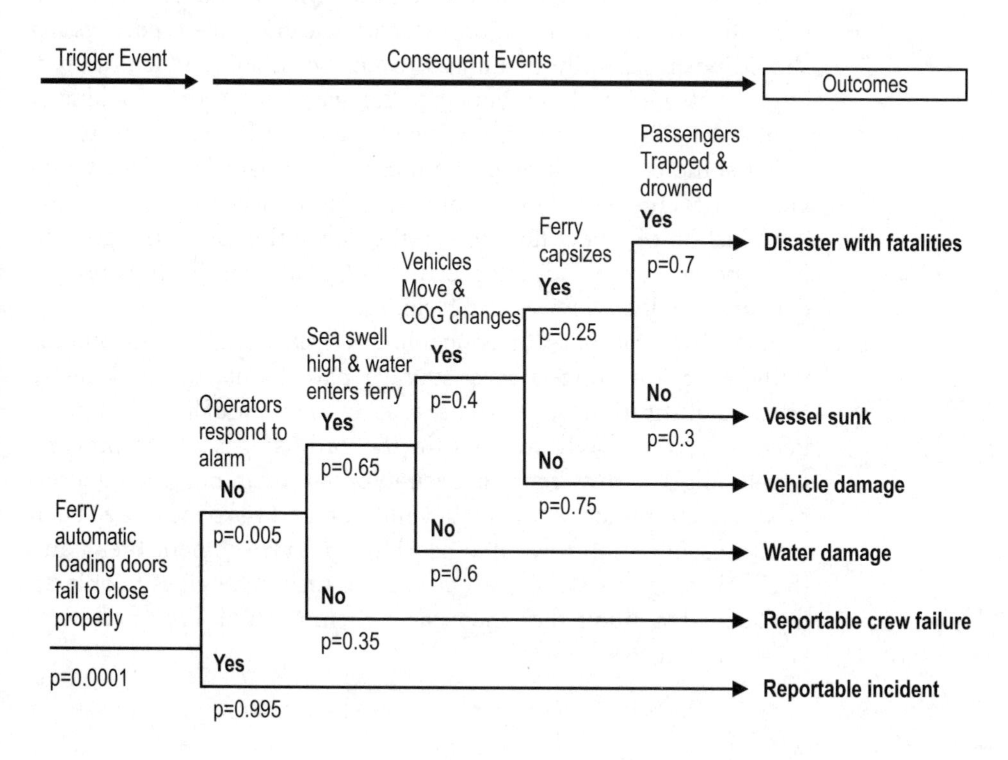

These in turn might become the means of defining strategic risk severity levels for the organisation. ETA techniques can be highly influential in strategic policy development within an organisation. Probability assessments can also be incorporated into ETA at the risk analysis stage.

SCENARIOS

Even where other techniques have already been used for risk identification on a project, it is usually worthwhile completing the process with a scenario-testing approach. Ideally, this takes a background of a particular historical or contemporary world view or event which has been the catalyst for significant change. From this background, a hypothetical scenario is established. For the chosen scenario, the question is posed: if there were to be a similar situation today (or if this type of event were to happen during the project), how would our involvement in the project be affected?

Examples of world scenario backgrounds might include the periods of economic depression in the 1930s, the political lead-up to the Second World War (or the Korean or Vietnam wars), the contemporary conflicts in the Middle East, the removal of the Berlin Wall, the breakdown of Soviet Russian hegemony in Eastern Europe, the release of Nelson Mandela from detention and the change to democratic government in South Africa, the Millennium and Y2K, or the September 2001 destruction of the New York World Trade Center. Obviously many, many more scenarios can be devised. Testing should not necessarily be limited to one scenario only. If time permits, a risk identification workshop can gain benefit from looking at more, particularly if they are framed to reflect different contexts.

For some projects, it might be more appropriate to replace world views or events with national or even regional scenarios, such as the introduction of a new tax system, the devolution of central government power, or the adoption of public/private partnerships for the delivery of public services and facilities.

Part of the purpose of concluding a risk identification process with scenario-testing is to encourage participants to break away from identification approaches that by now are likely to have become too narrowly focused. In the early stages of the process, developing a narrow focus is desirable, but it can become counterproductive if prolonged. After several hundred risks may have been identified in an intensive manner, having a refreshment break and then finishing this stage with a scenario technique gives people an intellectual stimulus that encourages them to think even more creatively about risks.

Risk identification resource documents

The documents used to record or reflect the myriad of decisions made in projects form a useful collective resource for risk identification. Typical among these documents are work breakdown structures (WBS), Gantt charts, activity networks, schedules, estimates, cost plans, detailed budgets, plans, drawings, specifications, performance standards and contract agreement clauses.

Official reports (e.g. the findings of accident enquiries) may also be used in risk identification. Their usefulness is usually limited to confirming the presence of particular types of risks, or the circumstances that are likely to give rise to them.

Outcomes and risk statements

The outcome of the risk identification process should be a formal record of the risks that participants have identified and are agreed upon as serious enough to warrant further investigation and treatment. Some risks may already be considered too trivial, even before the risk analysis stage, and there is little point in continuing to deal with them in the RMS, although it may be possible to gather several together and treat them as a group. The real benefit that will have been gained for the organisation through its risk identification is the assurance that the process has been effective; that either new risks have been identified, or known risks confirmed.

However thorough the process, it is unlikely that every project risk will have been identified at this stage, especially by an organisation unfamiliar with formal approaches to risk management. The inclusion of risk monitoring and control processes in the RMS cycle acts as a further precautionary measure against overlooking serious risks.

Identified risks should be recorded as formal, precise statements of likelihood, event, consequence and time. The greater the precision of each of these statements, the more straightforward and thorough the subsequent risk analysis process can be, and the easier it is to detect event/consequence contradictions or anomalies. Single word risk statements should be avoided. Their potential for inducing misunderstanding is so great that they constitute a substantial risk for the organisation themselves.

Typically, a complete risk statement will take the form:

There is a "p" chance that "w" event will occur during "x" period, with consequences "y_1" to "y_n" over the "z" period.

The most important variable to be stated is the event 'w', since the subsequent process of risk analysis will force attention to be given to the other variables. If the risk statement includes them, however, this will ensure that no risk variable is overlooked during the analysis.

An example of a risk identification statement for a construction project might be:

> **There is a chance that exceptionally wet weather will occur during the foundation excavation work, with the consequence that excavations will be flooded, progress will be delayed, and extra rectification costs will be incurred.**

Note how the progress delay and extra costs are correctly stated as the consequences, or impacts, of the natural weather risk. Note too that none of the risk statement variables are expressed as quantitative amounts. This is the work of the risk analysis stage.

6.7 CHAPTER SUMMARY

This chapter has presented the broad framework for a stakeholder-based project risk management system. It has been argued that project risks are really the risks of a particular stakeholder involved in a project, and that a single project-based RMS applicable to all stakeholders is not feasible, since different stakeholders will seek to fulfil different objectives, and will be engaged in different risk contexts, even for the same project. Each project stakeholder therefore needs to implement its own individual organisational RMS.

The procedural stages in risk management comprise risk identification, analysis, response, monitoring and controlling during the progress of a project, and the subsequent capture of risk related knowledge.

It is important to ensure that the appropriate context for risk management is understood clearly by everyone who is participating in the risk management process for an organisation on a particular project. In a similar way, risk management is informed by the objectives of the project and the objectives of the participating stakeholder. All these provide the essential guide to establishing the framework necessary to deal with the risks of specific projects and allow the risk management team to proceed with the exploratory stage of risk identification. Brainstorming is an inevitable ingredient of the many techniques available for identifying risks, and this is best undertaken in a risk workshop environment. Once risks have been identified, they can be analysed in greater detail. This is discussed in chapter 7.

CHAPTER 7

RISK ANALYSIS

7.1 INTRODUCTION

This chapter continues the exploration of systematic risk management, taking up from the identification of risks discussed in the previous chapter. The risk statements that are the outcome of the risk identification process form the input to the risk analysis stage of the RMS cycle. Risk analysis is an evaluative process that serves the purpose of establishing some understanding of the magnitude of the risks faced by an organisation in undertaking a project. The analytical process decomposes each risk into its constituent components and subjects them to some form of assessment. While the risk analysis stage itself is often lengthy, it is not always necessary for the risk assessments to be exhaustively exact in terms of mathematical accuracy. Indeed, accuracy criteria would be difficult to define in terms of many of the risk factors relating to projects. The financial performance of an investment project, such as a proposed shopping centre, for example, can rarely be forecast accurately (and perhaps more importantly, reliably) to within two or three percentage points in early stage modelling of internal rates of return for the project.

The nature of many projects also militates against high levels of accuracy in risk analysis assessments, since the input data are rarely derived from large sample statistics such as those used in the insurance industry or in large-scale manufacturing. For the most part, project data are collected from case-based small samples and gathered in unique situations. This also distinguishes them from data from repeatable experiments conducted under controlled laboratory conditions. However, what such case-based project data may lack in statistical adequacy is often made up by their capacity to generate information that is rich in content.

Appropriate risk analysis allows an organisation to gain an understanding of the relative severity of its risks on a project. In turn, this permits a strategic approach to managing them.

7.2 RISK ALLOCATION

Before embarking on the risk analysis process, a stakeholder organisation should investigate what prior risk allocation arrangements already exist for the project. There is little point in analysing a risk further if it has already been allocated to another party through a mechanism such as a special clause in a contract agreement. For example, a subcontractor normally used to submitting tenders on a fixed price basis would usually identify the chance of high economic inflation as a risk factor on a long-term project. However, if the head contract included clauses agreeing to share cost fluctuations between client and main contractor, with flow-on provisions for subcontractors, this would mean that the consequences of the inflation risk would be borne partly by the subcontractor and partly by the main contractor, who in turn would share the increased cost with the client. The contract agreement has effectively allocated the risk on a shared basis, and the subcontractor would need to reflect this in his assessment of the inflation risk.

In theory, of course, any risk should be allocated to the party best able to control it. Practice, however, especially in the guise of contracts, has a habit of mocking theory!

7.3 A THREE-DIMENSIONAL RISK MAGNITUDE PERSPECTIVE

Armed with knowledge of the prevailing risk allocation mechanisms of the project, the stakeholder can proceed to assess the likelihood of occurrence (chance or probability), consequence (impact), and time (duration) aspects of the risks that have been identified. Together these provide an important three-dimensional perspective of risk, as illustrated in figure 7.1, giving more substance to the concept of risk itself.

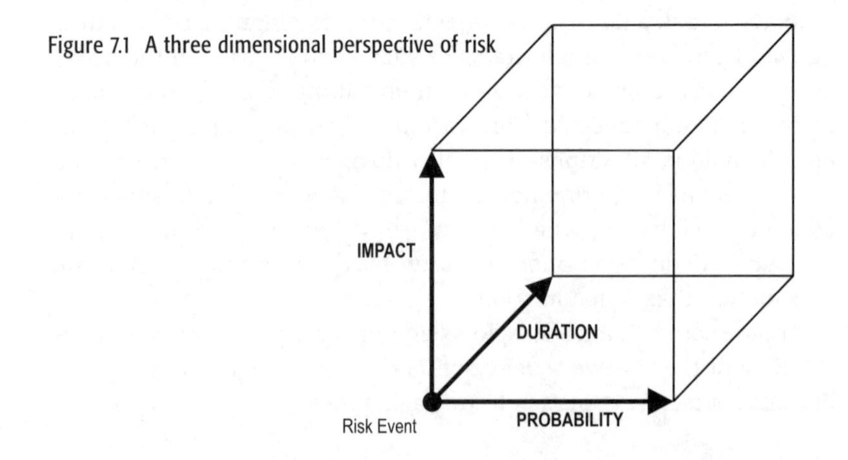

Figure 7.1 A three dimensional perspective of risk

Before considering each of these components of risk, the matter of uncertainty must be addressed once again, since uncertainty states can affect any or all of them.

Dealing with uncertainty

Risk analysis inevitably involves making estimates of variables that may lie far ahead in the future. Because the future cannot be known with certainty, uncertainty enters the scene. There will be uncertainty associated with the values of nearly all of the input variables from which the estimates are derived. Any one of a range of values might actually occur for a variable. Much of this view of uncertainty was discussed and illustrated in chapter 2, and considered again in chapter 3 where we looked at project complexity. Returning yet again to uncertainty in this chapter serves to emphasis its pervasive presence in projects.

Uncertainty is not necessarily confined to the values of input variables in an estimate. If the estimating technique is based upon some form of mathematical model, there may be uncertainty associated with the model itself. The model may not perform consistently under different conditions, nor with different ranges of input variable values. Thus, uncertainty of variable input values, plus model uncertainty, equals inherent variability in output values.

It should be possible to deal with model uncertainty by calibration, i.e. measuring the performance of the model over the whole range of possible conditions and comparing this with the real outcomes of what the model is trying to represent. The deviations from reality can then be translated into measures of confidence in the model. For example, it might be possible to describe a simple area/cost model for a particular type and configuration of building (i.e. a model which calculates the estimated construction cost for a proposed building by multiplying the total floor area by a price rate per m^2) as having a 95 per cent confidence level that the calculated cost will be accurate to within ±25 per cent of the real cost determined after the building has been built (assuming that the costs of any post-estimate change in project scope or quality are excluded from the post-completion cost calculation).

Several techniques exist for the treatment of uncertainty in the input values for risk analysis models. The principle that applies is that any method must be appropriate to the modelling context and logically defensible.

Quantitative approaches to treating uncertainty generally use mathematically based analyses of statistical data, and may involve simulation. At a simple level, the analysis might comprise sensitivity testing of the critical variable values: changing the value of one variable by one incremental step at a time and observing the effect on the outcome. The limitation of sensitivity testing is that it is not suitable for exploring the effect of multiple changes in several variables at the same time.

At a more sophisticated level, Monte Carlo simulation may be appropriate. In the introduction to this book we stated that it was not intended to be a mathematical treatise on probability, and simulation will not be discussed further. Many texts deal comprehensively with Monte Carlo and other forms of simulation, and it is worth noting that several computer software packages for Monte Carlo simulation are commercially available, either as independent software applications in their own right or as add-on modules for popular spreadsheet and project scheduling software.

The main criteria for using quantitative approaches for treating uncertainty in risk analysis are adequacy and sufficiency of variable input data, cost of data collection and cost of assessment. If the data are suitable and readily available, and if the assessment process is quick and straightforward, then quantitative approaches are preferable as they are objective. If sufficient data cannot be obtained, or if the cost/benefit ratio of collecting data or processing it is too high (and this may include techniques such as obtaining the consensus judgment of experts using Delphi methods), then qualitative approaches may have to be used. Earlier in this book it was noted that, in much of project risk analysis work, qualitative analysis is sufficient, since often the main purpose of the analysis is simply to gain an understanding of the relative severity of the risks that the stakeholder organisation faces in its involvement in the project.

In exploring the assessment of each of the components of risk, we will consider the necessary basis for each type of assessment and then present a qualitative model to represent it.

Qualitative assessment is subjective because it relies largely on human judgment. That judgment will be influenced by experience and by personal biases. Errors in human judgment include errors arising from biases such as:

- anchoring (wrongly sticking with a first estimate even though it is inappropriate)
- selective recall (remembering only certain facts or incidents)
- base rate fallacies
- over-pessimism
- over-optimism
- over-confidence in the assessment
- inappropriate search for patterns in data
- inappropriate framing of the situation.

The advantage of subjective assessment is that it is generally quick to apply and simple to understand. The disadvantage is that errors in judgment are often difficult to detect and eradicate. Because this is happening in human decision-making, the fields of decision science and behavioural psychology have much to offer in terms of addressing the problems of human judgment, and you are recommended to further reading in these areas. Reluctantly, they have to remain beyond the scope of this book.

Assessing the likelihood of occurrence

Probability, expected frequency, chance, and likelihood are used synonymously for this component of risk. The mathematical probability that a risk event will occur is expressed either as a percentage or as a decimal fraction greater than 0 (event certainly cannot occur) and less than 1 (event certainly will occur). The expression is really a contraction of a ratio relationship, so a 1 per cent (or 0.01) chance that a risk event will occur is really stating that there is one chance out of every 100 possible occasions that it will happen. There is a converse to the relationship, of course. If we believe that there is a 1 per cent probability of something happening, we must also believe that there is a 99 per cent probability of it not happening. This tells us that, for a finite number of possible and mutually exclusive alternatives for an event, the sum of the probability for each must equal 100 per cent (or 1). Thus, if the weather forecast indicates a 25 per cent chance of light rain showers over the next 24 hours, with a 5 per cent chance of heavy rain, then there must be a 70 per cent chance that no rain will occur during that period (assuming that light rain, heavy rain and no rain are the only possible and mutually exclusive alternatives).

The latter example brings in the issue of exposure period, since technically the likelihood of occurrence of an event must relate to some dimension of time. In some fields there is no problem about this. In surgery, for example, it might be stated that the 5-year survival rate for males between 50 and 55 years of age who have had prostate cancer surgery is 78 per cent (note how specific the context is here, and how the medical profession has used the optimistic converse: the death rate is 22 per cent). Other fields also use specific contexts and explicit periods: e.g. number of road accident deaths annually per 100 000 of population. Because of the probability convention used, in each example it is possible to make credible comparisons with another relevant context, such as female breast surgery or deaths due to industrial accidents.

Projects do not really 'work' in the same way. Potentially each has multiple environments (procurement, operation, and disposal); up to four elements (tasks, technologies, resources and organisation); and usually a great number of differentiated and interdependent subelements. The many hundreds of identifiable risks for a project stakeholder are therefore likely to exhibit a great array of different exposure periods, and most of these may not be known precisely. This, together with the difficulties of obtaining reliable probabilistic data, is why the assessment of risks in project risk management tends to be undertaken using qualitative approaches. For the most part, the 'time' aspect of risk event probability is either ignored, or treated implicitly as equivalent to the project procurement period. Later in this chapter you will see that we have chosen to give this issue more explicit treatment, although still qualitative to a large extent.

A useful guide to making qualitative expressions of probability or likelihood is given in AS/NZS 4360 (1999). This risk management standard offers a 5-interval scale of linguistic descriptors for the likelihood of occurrence for a risk event, as shown in table 7.1. It is immediately apparent that the scale is generic, as it does not relate to any specific project context. Indeed, the whole of AS/NZS is generic in that it is not uniquely project focused. Nor is the scale specific as to time.

Table 7.1 Interval descriptors for likelihood

Interval descriptor	Amplification
1. Rare	May only occur in exceptional circumstances
2. Unlikely	Could occur in certain circumstances
3. Possible	Might occur at some time
4. Likely	Likely to occur in most circumstances
5. Almost certain	Expected to occur in most circumstances

Source: adapted from AS/NZS 4360 (1999).

One implication of offering a generic scale is that, before using a qualitative instrument such as this, a stakeholder organisation must establish interpretations for the scale intervals that are meaningful in terms of its involvement in, and objectives for, a project. For most projects there is no point in all the stakeholders coming together and agreeing uniform meanings for each interval – there will be too much disparity between the nature of their individual involvement and the objectives each seeks to achieve.

A further implication that arises is that, where qualitative approaches to risk assessment have been used, any communication about risk between stakeholders will be affected. If different stakeholders maintain different interpretations of 'rare' or 'likely', for example, they are unlikely to share completely congruent understanding.

On the other hand, it is desirable and practical for an organisation to decide on uniform meanings for its scale interval descriptors across the organisation, which can then be applied to all the projects it undertakes. Without such uniformity, an organisational risk register would lose much of its usefulness.

Given an agreed scale, such as that in table 7.1, a stakeholder organisation has the opportunity to 'score' its qualitative assessments of probability for identified project risks. The results can then be incorporated into a risk severity model as shown later in this chapter.

Assessing the consequences

A risk event that occurs on a project must have consequences, since this is fundamental to the definition of risk. From the 'dual' view of risk, these impacts may be positive or negative; beneficial or adverse, but it must be remembered that, for the time being, risk events in this book are presented as leading to negative consequences.

The consequences of a risk event will impact the project, but more importantly they impact the stakeholder bearing that risk. Impacts from the same event may also be experienced by other stakeholders in the project and, in some instances, by others beyond it.

For the purposes of assessing the consequences of a risk event, it is often assumed that the impacts can be expressed in terms of cost to the risk bearer. Where quantitative assessment of probability has been used, this assumption is even more strongly held, since it allows the assessor to calculate a financial exposure to the risk, and thus a total financial exposure to all risks. Thus if a risk event has a 5 per cent chance of occurring each year, and the estimated consequence is $1000, then the risk bearer is said to be exposed to $50 of risk ($0.05 \times \$1000 = \$50$) and might wish to consider spending up to $50 a year to eliminate or avoid it.

One problem with this approach is that not all risk consequences are directly related to costs. Among many others, the outcomes may lead to impaired reputation, vulnerability to legal prosecution, loss of capacity, loss of staff, difficulty in recruiting staff, lowered resilience, etc. Assessing monetary values for any of these, while it can be done (and is done by the courts, for example), could be extremely difficult for an organisation unfamiliar with such a task. It would almost certainly involve some degree of financial gymnastics.

For a qualitative approach, AS/NZS 4360 (1999) again offers guidance. The 5-point interval descriptor scale for risk impacts in the risk management standard is shown in table 7.2. With this scale, AS/NZS 4360 has tried to denote more specific contextual impacts by referring to 'containment' and 'toxic release', but these should not be seen as limiting the applicability of the scale. It also refers to financial loss and we have already noted the difficulties that can arise with assessment of this.

As with the qualitative interval descriptor scale for likelihood, a stakeholder organisation must assign interpretations to the scale

Table 7.2 Interval descriptors for risk impacts

Interval descriptor	Amplification
1. Insignificant	No injuries, low financial loss
2. Minor	Minor injuries, rapid containment on-site, medium financial loss
3. Moderate	Medical alert and light injuries, on-site containment requires outside assistance, high financial loss
4. Major	Extensive injuries, loss of production capability, off-site release with no detrimental effect, major financial loss
5. Catastrophic	Deaths occur, toxic off-site release with detrimental effect, huge financial loss

Source: adapted from AS/NZS 4360 (1999).

intervals that are internally meaningful. For example, it may be appropriate for the organisation to assess the impacts of risk events in terms of their capacity to delay the project. Scale interval descriptors such as 'not greater than 2 hours; between 2 and 6 hours; 6–12 hours; 12–48 hours; and greater than 48 hours' might be appropriate for organisations used to a very short-term involvement with projects, or for event-type projects with an extremely short duration. Actual monetary amounts could be set against the scale intervals where this is practicable, and in fact any scalable impact of risks borne by the organisation could be treated in a similar way.

Care is needed in setting the interpretive levels. An organisation in the early stages of RMS maturity (this is discussed more fully in chapter 9) will tend to score too many of its risk impacts too highly simply because it has not set the bar high enough for the upper level descriptors. A 'catastrophic' impact descriptor, for example, should be reserved for a risk impact that might completely destroy a company financially and cause it to cease to exist. While none of the interval scales offered by AS/NZS 4360 is intended to convey absolute value meanings, and while the scales should therefore not be used to attempt to calculate precise impact values, their purpose is to permit some meaningful comparison of the relative severity of risks. The interval descriptor interpretations that an organisation assigns to the scales should therefore have some reasonable correspondence with reality for that stakeholder.

The impact assessment 'score' for each identified risk may also be incorporated into a simple risk severity model.

Assessing the duration of exposure

Earlier in this chapter we noted that, strictly speaking, the duration of exposure belongs to the component of risk that relates to the likelihood of occurrence (the chance or probability) of the risk event. It was suggested that, because of the nature of many projects, the risks that arise from them do not readily fit this time perspective neatly, at least not for the purposes of assessment by a project stakeholder. This is because the risk event may be framed by one time window, and the consequences by another. The risk event might occur in one project environment, but its impact could be felt in another.

For example, in most projects there is a chance that poor quality workmanship could occur during the procurement phase, but the consequences might not become evident until well into the operational phase, or even not until the disposal phase. This drawing out of exposure time, perhaps for as much as 30 years or more, is a complicating issue for risk management. It is usually the reason why time is ignored in much of risk analysis except for discounted cash-flow models for investment projects and some quantitative long-term health project studies.

While the time factor may complicate quantitative risk analysis, it can be addressed quite simply in qualitative analysis. For convenience, a 5-point interval descriptor scale can be devised to suit the characteristics of the project types most familiar to the stakeholder organisation in its field of operations. As before, each stakeholder must assign meaningful values to the scale interval descriptors. The indicators described in table 7.3 are therefore only suggested interpretations and should be tailored to suit individual circumstances. For some stakeholders dealing with particular projects, long-term risk exposure durations might last no more than a few weeks. For other stakeholders involved in different projects they could endure for many years. A stakeholder might experience risk exposures spanning only the procurement environment; but another might be vulnerable throughout the operational and disposal environments as well. The time value assigned to each risk should include the period of exposure to the risk event and the period during which the consequences might flow.

Table 7.3 Interval descriptors for risk duration (exposure time)

Interval descriptor	Possible indicator
1. Short term	Less than the time required for procurement phase
2. Medium-short term	Procurement phase plus one third of operational phase where appropriate
3. Medium term	Procurement phase plus half of operational phase where appropriate
4. Medium-long term	Procurement phase plus two thirds of operational phase where appropriate
5. Long term	Into disposal phase and beyond

Armed with these qualitative approaches to assessing the likelihood of occurrence, impact and duration of exposure for project risks, it is now possible to consider combining them into a subjective risk severity assessment model.

Combining probability, impact and duration

Given the three-dimensional concept of risk presented by figure 7.1, and the rating scales discussed above (tables 7.1, 7.2 and 7.3) for the probability, impact and exposure duration components of risk, combining them into a simple risk severity model is quite straightforward.

Many approaches to qualitative risk analysis suggest two-dimensional models, using 3- or 5-point interval scales for probability and impact only. Scoring each risk is simply matter of multiplying the chosen probability rating for the risk by the chosen impact rating. With maximum risk severity scores of 9, 15 or 25 (depending upon the scales used), the results may be too coarse to allow more subtle distinctions between different risks. There will be too many with identical scores. It would be possible to expand the rating scales to perhaps 9 or 11 point intervals, but this would pose a difficult linguistic problem in assigning realistic descriptor labels to each interval, to reflect sufficiently finer shades of meaning. In any case, the main drawback with two-dimensional scoring models is that they ignore the important dimension of time. A three-dimensional risk severity model, based upon 5-point subjective rating scales for probability, impact and duration, is shown in figure 7.2.

Figure 7.2 A subjective three-dimensional risk severity model

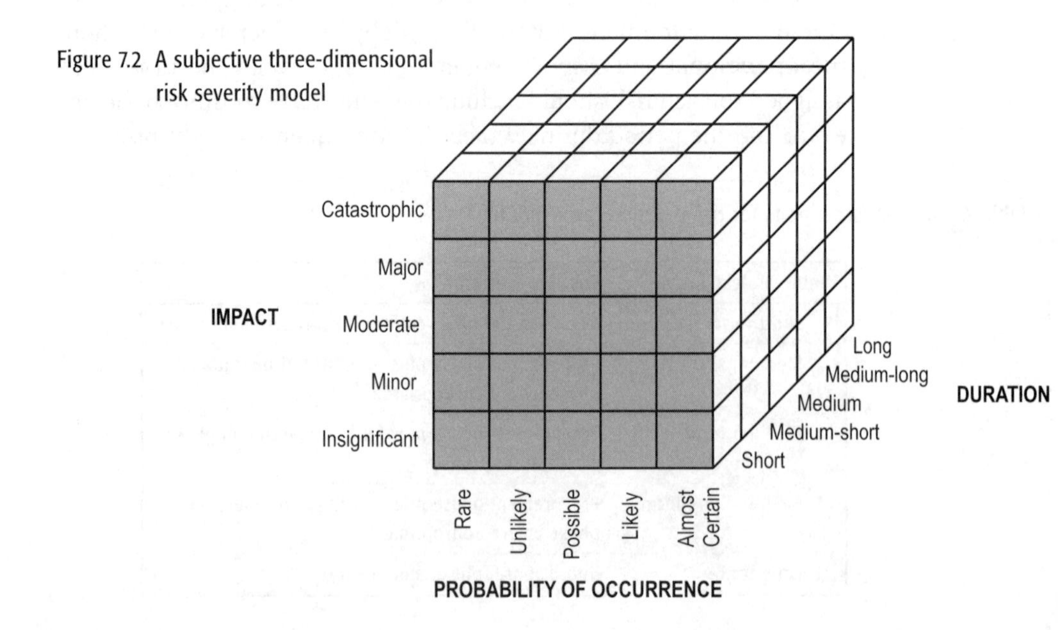

Using the model to assess a risk is simply a matter of assigning a value between 1 and 5 to each of the three scales. This will produce a potential risk severity score ranging from 1 ($1 \times 1 \times 1$) to 125 ($5 \times 5 \times 5$). Theoretically, finer-grained scores could be obtained by assigning decimal values to any scale (e.g. 0.75 probability $\times$ 1.5 impact $\times$ 3.25 duration = 3.65625), but in practice this is unnecessary and integer values are sufficient. Injecting decimal precision into the scale values is more often than not counter-productive in terms of the extra effort required, and begins to defeat the objective of the assessment being done at this stage. The purpose of the model is simply to establish some measure of comparative severity for the risks that a stakeholder organisation faces in a project. It is not intended to produce absolute results. The severity scores produced by the $5 \times 5 \times 5$ model allow the organisation to confidently rank the identified risks and hence to prioritise the treatment options and decisions for them. The most severe risks will have to be revisited for further, more detailed analysis anyway, so there is little point in prolonging the time needed for initial assessment by setting unnecessarily high precision targets for the subjective scales of the model.

7.4 OTHER ASSESSMENT TECHNIQUES

Other risk analysis and assessment techniques are available which can be used either objectively or subjectively, according to circumstances. Prominent among these are 'expected utility' and 'expected monetary value'.

Expected utility

The expected utility (EU) assessment technique originates in decision science, and is simply the product of probability and impact. This expression of risk was noted earlier when the assessment of risk consequences was considered. EU assessment may be applicable when the direct and indirect financial impacts of risk events are difficult (or even impossible) to estimate. 'Utility' in this case is some measure of the loss of worth, or loss of usefulness, to the risk taker if the risk event should happen. An example, using a decision tree approach, will help to demonstrate the technique.

Example 7.1 uses a travel itinerary project as a scenario. From the market survey and time performance data that the tour company has obtained, it is able to draw up a decision tree as shown in figure 7.3. The decision nodes are shown as A (flying), B (bus) and C (train). Since the probabilities of arrival on time at the destination town are known from the statistical performance data the company has obtained, the chance of delay for each travel mode can be determined

Example 7.1

TRAVEL TOUR PROJECT

A tourism company is investigating a project to introduce a new short tour package for foreign customers. The package will include an inbound flight to a national capital city. After a few days sightseeing in the city, the tourists will travel to another major town which has internationally renowned shopping precincts, before flying home again two days later.

The tour company has narrowed the feasible travel options between the two towns to three: a short domestic flight lasting about an hour; an eight-hour luxury coach trip; or an eleven-hour overnight journey in a train with sleeper facilities.

Price is not a factor in the decision, since the total cost of the tour package will be about the same whichever option is chosen.

The company has carried out a market survey of potential customers for the tour package. Using a carefully worded questionnaire, it has established 'best' and 'worst' case customer impact outcomes for each of the inland travel options. Survey respondents voted for air travel as their most preferred choice, provided the flight left promptly and arrived on time. The rail journey was considered the next best, since the overnight sleeper was thought to be quite exciting, and the railway station at the destination was located in the town centre. Again, delays were frowned upon. The least preferred option was the bus journey. Even though a luxury coach would be provided, survey respondents thought it would be too cramped and uncomfortable over the eight-hour trip.

At substantial cost, the tour company also managed to obtain performance data on flight, bus and rail travel between the two towns. These were sufficient to permit calculation of the likelihood of arrival on time for each of the three modes of travel.

Figure 7.3 Decision tree and expected utility

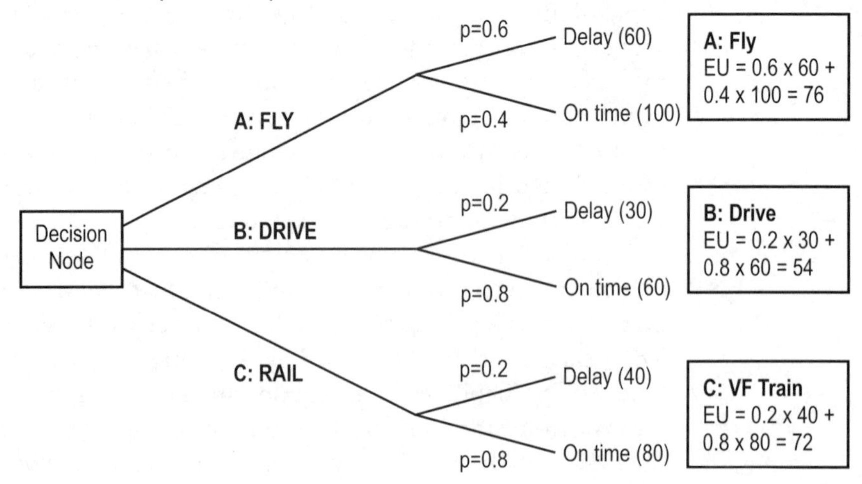

(100 per cent – x per cent chance of timely arrival). The values in brackets at the end of each paired branch, assigned to arrivals on time and to delays, were derived from the responses to the carefully worded market survey questions as these could be translated into the perceived worth of each outcome to the potential customers.

The EU calculation for each travel option is shown in the small boxes on the right-hand side of the diagram. The resulting EU value is the sum of the product of probability and utility for the paired 'on time'/ 'delay' arrival condition states for each travel option. It represents the average worth to the traveller for that option had the journey been made, and both condition states experienced, on many occasions. Theoretically, the travel option yielding the highest utility (flying: EU = 76) should be selected for the itinerary, as it represents the alternative with the least risk of causing dissatisfaction to the tour company's customers.

Some of the difficulties associated with this technique are obvious. Obtaining objective time performance data for the aircraft, buses and trains would probably be the most straightforward task, but even that would not be simple. The data would be expensive and time consuming to gather, and almost certainly some degree of data manipulation would be required. The market survey instrument would be expensive to design and administer, and the response data would have to be carefully processed and interpreted. Significance testing would be necessary and the results of the model should be compared with those from other decision-making approaches to confirm the validity of the EU technique. More realistically, perhaps, would one reasonably expect a tour company to make its project decisions in this way?

Expected monetary value

Expected monetary value (EMV) is simply a financial version of EU, and used in situations where reasonably precise estimates of various possible financial outcomes can be made, as well as the probabilities associated with their occurrence. Example 7.2 demonstrates this.

The organisation now has the data it needs to calculate an EMV for a projected loss in the first year of trading if it purchases the franchise. Table 7.4 shows the detailed calculation. Note that the available data reflect the situation for a \$100 000 franchise purchase, while the micro-business is considering an investment of \$50 000. In the absence of any other information, it is reasonable to make pro-rata adjustments to the data as shown, but the organisation should be aware that greater uncertainty will be associated with the adjusted data.

Example 7.2

GARDENING FRANCHISE PROJECT

A micro-business organisation is exploring the feasibility of purchasing a well-known gardening franchise for $50 000.

The company has access to information from the National Association of Business Franchisees (NABF) about typical business performance in franchising operations. According to NAF, the majority of new franchisees suffer financial losses in their first year of operation. However, many then recover and trade successfully from there on. The research data suggests that, for a $100 000 franchise purchase:

1. There is a 50 per cent chance that the loss in the first year would not exceed $15 000.
2. There is a 25 per cent chance that the loss in the first year would not exceed $10 000.
3. There is a 15 per cent chance that the loss in the first year would not exceed $7 000.
4. There is a 10 per cent chance that the loss in the first year would not exceed $5 000.

Table 7.4 EMV calculation for first-year loss in gardening franchise

Probability (A)	$ Loss (B)	A x B = $C	$C x 50%
50%	15 000	7 500	3 750
25%	10 000	2 500	1 250
15%	7 000	1 050	525
10%	5 000	500	250
		EMV =	$5 775

The EMV loss of $5775 in the first year of trading is what the organisation should realistically plan for if it decides to purchase the franchise. If a more conservative business plan is preferred, it could assume a $7500 loss.

As with the EU technique, there must be suitable data available – and the user must have confidence in their accuracy and reliability – for the EMV to be considered as a practical objective risk assessment tool. Otherwise, it is no better than any other subjective assessment technique.

7.5 RISK RANKING

Once identified risks have been analysed and scored in some way, they can be ranked in terms of relative severity. This is a useful way of focusing the stakeholder organisation's attention on the most serious risks it faces on a project.

While a straight ranking of risks according to their severity scores is useful, an organisation could adopt a more strategic approach by aligning these scores to a set of predetermined risk severity categories. A five-category list, adapted from AS/NZS 4360 (1999) is shown in table 7.5.

Table 7.5 Interval descriptors for risk severity

Interval descriptor
Minimal
Low
Moderate
High
Extreme

Source: adapted from AS/NZS 4360 (1999).

Each stakeholder organisation should decide upon how to align its 125-point severity scale to the five-category list, but it is important to caution that an equidistant interval alignment is not advisable (i.e. assigning 'minimal' category to risks with scores between 1 and 25; 'low' risks between 26 and 50, etc.). The use of strategic risk severity categories is discussed in chapter 8.

Once risks have been ranked and prioritised, it is likely that the most severe risks will have to be revisited for further analysis before major decisions about their treatment can be taken. Eventually however, a response must be considered for each identified risk.

7.6 CHAPTER SUMMARY

While, by intention, this chapter has not dealt with highly mathematical and financial modelling approaches to risk analysis (and each of these warrants a separate book in its own right), many techniques of risk analysis have been presented, ranging from semi-quantitative to subjective and linguistic assessment tools. The treatment of each of these has been brief rather than extensive. In real life the risk analysis stage of formal risk management can be quite time consuming. Each of the components of risk – the likelihood of occurrence of the risk event, the consequential impact, and the duration of exposure – must be considered carefully for every identified risk that the stakeholder organisation faces on a project. In this assessment of risk severity, an organisation must select and use techniques appropriate to its risk evaluation objectives.

Once risks have been sufficiently analysed, it is possible to make informed decisions for dealing with them.

CHAPTER 8

RISK DECISIONS AND ACTIONS

8.1 INTRODUCTION

Up to this point, the exploratory and evaluative processes of formal systematic risk management have been treated in a somewhat passive, abstract manner. Even in real-life projects, it is often found that risk identification and risk analysis create a similar impression on people engaged in these tasks: they feel distanced and separated from the risks themselves.

While the discussion in this chapter will continue to be dispassionate, real risk management protagonists working for real stakeholder organisations on real projects are likely to experience much more direct engagement at this stage of the risk management process. This is because the time has come to make decisions about the risks facing the organisation, to implement those decisions and to take actions for monitoring and controlling their outcomes during the progress of the project. The topics covered in this chapter therefore deal with risk response, risk monitoring and control, risk recovery, and risk knowledge capture.

8.2 RISK RESPONSE OPTIONS

There are four basic responses to risk: avoidance; transfer; reduce and retain residual; and retain. The first option is exclusive, since if a risk is avoided it cannot be transferred, reduced or retained. The remaining options, however, are frequently encountered in combination.

Risk avoidance

Avoiding a risk means deliberately taking another course of action so that it cannot arise in the new circumstances for that project. Note that this is not the same as eliminating risk. In fact, risk elimination is rarely possible: if it becomes necessary to revert to the former circumstances the risk will return.

The ultimate form of risk avoidance is not to proceed with the project at all. However, this extreme is seldom adopted as a response to one project risk factor in isolation. Such a decision is usually arrived at after a number of issues have been considered and their overall influence on the project assessed. An investment project opportunity, for example, is not rejected simply because it exhibits a potential rate of return which is not commensurate with the investor's criterion rate, but because improvements in performance cannot readily be achieved in the major factors that contribute to the rate of return calculation.

In exploring risk avoidance, a stakeholder organisation should always keep in mind that this response might thereby increase the severity of other risks, or result in lost opportunity. An example serves to illustrate this.

Assume that a construction project, such as a multi-storey residential building, has been designed with basement parking. However, the site has a history of occasional flooding following periods of exceptionally heavy rain. The risk is the chance that a heavy rain storm will flood the basement, causing damage to the vehicles parked there and inconvenience to the residents. In this case, a risk avoidance measure would be to redesign the building to omit the basement levels and replace them with ground or above-ground level parking. The risk of the basement becoming flooded is avoided since there is no longer a basement in the proposed building. Instead, however, the risk of flooding at ground level is increased – although the impact of this might not be as serious. Furthermore, the design change might alter the visual effect of the completed building sufficiently to affect the selling or letting market opportunities for the residential units. Finally, of course, the cost of making the design change (in terms of the construction costs, the design costs, and any approval or delay costs incurred) have to be weighed against the potential impact costs of the risk avoided.

Risk avoidance is rarely a straightforward or easy response measure. Because it is highly effective in the right circumstances, however, it should always be considered first among the possible options available to a stakeholder.

Risk transfer

In electing to transfer risk, a project organisation is seeking to shift the burden of a particular risk to another stakeholder. This is a common response in project situations where stakeholder supply chains or networks are easily distinguished, as attempts will be made to transfer risks progressively along the supply chain or to the more distant extremities of the network. Typically, a project client will transfer risks to a contractor, who in turn will transfer them to subcontractors or suppliers.

The mechanisms used to transfer risks in such situations include, *inter alia*, head contract agreement clauses, subcontracts and supplier agreements.

The principle of risk transfer forms the distinguishing characteristic between many alternative forms of project procurement and delivery, such as joint ventures, public/private partnerships, and the Private Finance Initiative in the United Kingdom. In the construction industry it is noticeable in procurement systems such as design/build (D & B), design-build-finance-operate (DBFO), build-own-operate (BOO), build-own-operate-transfer (BOOT) and many others. Common to nearly all of these is the intention to transfer risk away from the client stakeholder and towards the contractor organisation.

Insurance is a transfer mechanism for risks that are insurable: theft, injury, damage to property or equipment. This introduces a third party risk transferee to the situation, thereby involving another stakeholder in the project. Performance bonds, sureties and payment guarantees are used in a similar way to deal with the impacts of the risk of default by parties in the execution of project contracts and agreements. The risk transferees may be banks, finance companies or other third parties.

Two things should be noted about risk transfer: it is usually costly; and it is rarely 100 per cent effective.

Every risk transfer has a cost. This is usually reflected in the price the transferee charges for accepting the risk - the premium payable to an insurance company, the fee set by a bonding agent, or the commission charged by the bank. These are fairly obvious forms of direct costs of risk transfer. More indirect costs occur when, for example, a risk is passed to a project subcontractor who does not manage it effectively and is then affected by the impact of a risk event. If the subcontractor organisation survives the risk impact, but cannot recoup its losses from the main contractor, it will attempt to do so on other future projects. If the subcontractor does not survive the risk impact, the losses are eventually borne by the players in the relevant industry and finally by society itself.

Few risk transfer mechanisms completely discharge the transferor from all responsibility associated with risks. Contemporary views of corporate ethics and social responsibility do not allow organisations to absolve themselves completely from such risks simply by transferring them to other parties. Transfer of liability for the consequences of risks events may be achieved, but accountability often remains in some degree. On a personal level, one cannot insure possessions against theft, neglect to take any precautions against theft occurring, and still expect to receive full recompense from the insurer if something valuable is stolen. In a similar way, a client stakeholder might transfer the risks of accident and injury involving project workers or third parties to the project contractor. However, should a serious accident occur, the client stakeholder will inevitably become involved in the resulting adverse media publicity, simply by reason of its association with the project. A project sponsor might transfer the risk of environmental damage to a contractor, but then find itself (even in other areas of its activities) the target of unwelcome public protest demonstrations.

Risk reduction and residual retention

No risk should be avoided, transferred or retained without first checking to see if it is possible to reduce it and then retain the residual risk.

For anyone with a professional interest in risk management, reducing risk is probably the most absorbing area of involvement. The stakeholder is deliberately attempting to minimise risk in some way. All dimensions of the risk should be examined, since it may be possible to reduce the probability of occurrence, the impact consequences, or the duration of exposure to the risk. Combinations of any two, or even all three, of these may be contemplated in some risk circumstances. The nature of the risk will influence which of the risk factors can be mitigated, and the context will influence the possible extent of mitigation. For natural risks, for example, it is rarely possible to reduce the chance of occurrence (although planning a project to avoid seasonal weather extremes is one approach).

The risk reduction process also allows the value of a risk classification system to be exploited. Whenever a new project risk is encountered, it is possible that an effective treatment can be found among more familiar risks (i.e. from the organisation's risk register) in the same category.

Inevitably, the process of exploring risk reduction requires a return to the analytical processes of risk assessment. Care must be taken in doing this. There is a natural inclination to try to replace earlier subjective qualitative assessments with techniques that are more quantitative. Provided adequate and sufficiently reliable data are available or readily accessible, this is acceptable. If better data are not available, then it is better to retain a qualitative approach since by now any subjective judgements involved should be better informed about each risk.

Attempts by project stakeholder organisations to reduce the likelihood of occurrence (chance, probability) of a risk event tend to focus upon implementing, or increasing the effectiveness of, other management techniques for the project. These might include: more stringent safety precautions (occupational health and safety management); more frequent financial audits (financial management); more frequent stock checks (logistics and asset management); more stringent or more frequent inspection of finished work (quality management); and regular or advanced training for staff (human resource management). On a construction project site, for example, the responsible safety officer might issue instructions for any temporary scaffolding to the outside of the building to be fitted with additional horizontal intermediate rails at each floor level, and with higher toe boards on the outside edge of the planked walkways. In doing so, he or she is trying to minimise the chance that either a worker or an object will fall from the scaffold. Both such incidents are among the most frequent causes of serious accidents in the construction industry worldwide. Note that the safety officer is addressing only the likelihood of occurrence of the risk event. If someone or something does fall from the scaffolding, the impact will be the same as though no extra precautions were taken. Suitably strong catch nets strung from the scaffolding would be one way of reducing the impact of an accident, but note too that this alone would not minimise the chance of the accident happening. Nor do any of the additional precautions affect the duration of exposure to the accident risk, as the scaffolding remains in place for the same period of time. In fact, for this example none of the precautions suggested so far actually addresses the root *causes* of scaffolding accidents. Prevention measures, such as improved safety training and safety motivation for workers, are likely to be better ways of achieving this.

In another construction example, it is sometimes necessary to disturb or remove the foundations of an existing building, in order to deepen them or replace them with new foundations (perhaps for a new adjoining, heavier building). This operation is known as underpinning, and is regarded as a risky process since it involves the temporary removal of structural support from the existing building. The risk reduction response in this case is to carry out the work in very small sections at a time, each of which is executed in a strictly planned sequence so that no two adjoining sections in the linear length of the existing foundation are exposed or removed at the same time. This underpinning technique reduces the likelihood that a foundation collapse will occur.

A principle detectable from these examples is that reducing the likelihood of occurrence of a risk event generally means looking for ways of doing things better, or doing them differently. Value engineering techniques can be useful here, since these are aimed at exploring alternative solutions to delivering required functions.

Reducing the impact of a risk event calls for a different approach. Risk impact reduction usually depends upon the prudent provision of extra resources, so that the potential consequences of a risk event are diluted in some way. The old financial investment adage about 'not keeping all your eggs in one basket' is a typical example of risk impact reduction. At one extreme, risk impact reduction precautions might amount to having additional sources of supply readily available for critically essential project materials. At the other extreme, large sums of money might be kept in reserve to pay off aggrieved people opposed to a project. The tools of risk reduction, therefore, are most often found in precautionary measures such as back-up resources and recovery planning, contingency allocations, reserve funds, public relations expertise, and portfolio investment.

For a theatre performance project, the director will appoint understudies for the leading roles. Doing this does not reduce the likelihood that a leading actor will become indisposed for a particular performance (although that may not be strictly true, since the motivating effects of having someone treading closely in your footsteps are well known in the performing arts!). The extra rehearsal payments to the understudy, and the costs of advising patrons and offering refunds to a few disgruntled ticket-holders, are the risk premiums for averting the possibility of having to cancel the performance completely and thereby losing significant box-office revenue.

Trying to reduce to the period of exposure to a risk can be a complicated process, since it is almost always necessary to consider the exposure to the risk event separately from the exposure to its consequences. The tool most appropriate for the former is rescheduling; while effective instruments for the latter may be the implementation of shorter periods for guarantees and warranties offered to customers. The one seeks to accelerate progress, and the other to minimise a period of liability.

In a book publishing project, the publisher might decide to increase the frequency and range of marketing activities during a given period for a particular book. In doing so, the publisher is trying to reduce the chance of a publishing failure, but more specifically the time during which his or her investment will be at risk. The additional marketing and publicity activities are intended to generate more sales more quickly, so that the sales break-even point can be reached more quickly. The trade-off is the extra cost of the additional activities.

Risk reduction processes may go on iteratively until the point is reached where the residual amount of a risk is acceptable and can be retained by the project stakeholder organisation. This means that more than one return to the risk analysis stage may be required during the decision-testing period.

Risk retention

Retaining risks without mitigating them presumes that the decision is an informed one and based upon analysis which indicates that any reduction treatment has a negative cost/benefit ratio. Retaining residual risks shares the same presumption, in terms of further reduction beyond that already achieved.

Beyond this, of course, there will be risks unwittingly retained by the organisation simply because they have not been identified.

In some instances it is possible for a stakeholder organisation to reward itself for retaining a risk. This presupposes that the risk identification and analysis processes of risk management have been implemented, at least to some extent. An excavation subcontractor tendering for a construction project might retain the risk of storm water flooding the excavations, but reward himself by increasing the unit price rate for the work. A developer might increase the 'hurdle' or required rate of financial return for a project as a reward for retaining a variety of risks, such as the economic risk of bringing the completed project to market in an unstable commercial climate. A financier will increase the loan interest rate as a reward for retaining the risk of payment default on a loan.

In all of these situations, the risk itself may be unchanged in terms of probability, impact and duration. The 'risk and reward' approach, however, must be carefully balanced as it introduces its own additional risks. The excavation subcontractor who includes a risk premium might find that his bid is unsuccessful because competitors have submitted lower prices. The developer might find competing projects selling (or letting) at lower prices as their developers are satisfied with smaller returns. The financier might find her share of the market diminishing as borrowers seek alternative loan sources at lower rates.

While the principle of risk retention is that it should be done on an informed basis, the principle of seeking compensatory reward should always be on the basis that the stakeholder is not thereby exposed to other, more severe, risks.

Combination responses to risk

Combinations of retention, reduction and transfer responses to risk are possible. Since risk avoidance aims to change the circumstances through which a particular risk arises, this response cannot be used in combination with others.

Probably the most common example of combination risk response is the transfer of risk through insurance, while at the same time retaining a small amount of the impact by accepting liability for a fixed excess sum in the insurance policy agreement.

Another example can be found in the 'target cost' variation of the cost-plus type of procurement system sometimes used for construction projects. In contracts for these projects, the risk of cost overrun is sometimes shared between client and contractor, as an incentive for the client to avoid scope changes and for the contractor to work efficiently. Contract clauses will make the sharing arrangement explicit (e.g. 50/50 or some other proportion). Co-incidentally, similar clauses might deal with the sharing of cost savings on the project, but this does not fall within our definition of a risk since it is not an adverse event (although the contractor's potential loss of anticipated profit on any cost saving might be regarded as a risk). A slight variation on the insurance example is where the risk transferee (in this case the insurance company) requires the transferor to carry out mitigating action as well as imposing an excess liability. For burglary insurance, an insurance company may require the policy-holder to fit special window and door locks, and install an alarm system. These precautions reduce the probability of the unlawful entry occurring. Risk transfer and retention is therefore combined with risk reduction.

8.3 STRATEGIC RISK RESPONSE

Where a project stakeholder organisation has established severity categories for the risks it faces (see table 7.5), it is possible to formulate strategic risk response policies as a guide to determining the level and type of management responsibility required. Such an approach is shown in table 8.1.

Table 8.1 Interval and strategic management descriptors for risk severity

Interval descriptor	Strategic management indicator
Minimal	Manage by exception flagging only
Low	Manage by routine procedures
Moderate	Specify management responsibility level and periodic attention
High	Requires frequent senior management attention
Extreme	Immediate and/or continuous action required; highest level of organisational responsibility assigned

Source: adapted from AS/NZS 4360 (1999).

The ALARP principle can be used in conjunction with a strategic approach to risk response. In terms of this principle, the stakeholder attempts to set criteria for its risk response decisions that ensure that only the least severe risks are retained. The more risk averse the organisation, the lower it will set the risk severity score or severity indicators for the risks it is prepared to retain. Setting these at the lowest level is not always feasible, hence the ALARP acronym (as low as reasonably practical). This concept is illustrated for a 125-point risk severity scoring scale in figure 8.1. In accordance with our earlier discussion of risk ranking, the strategic scale intervals should not necessarily be spaced equidistantly. For a risk-seeking organisation looking to exploit project opportunities, the converse principle AHAPM (as high as possibly manageable) might be applied.

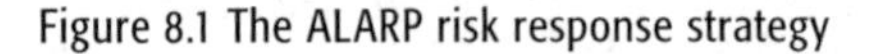

Figure 8.1 The ALARP risk response strategy

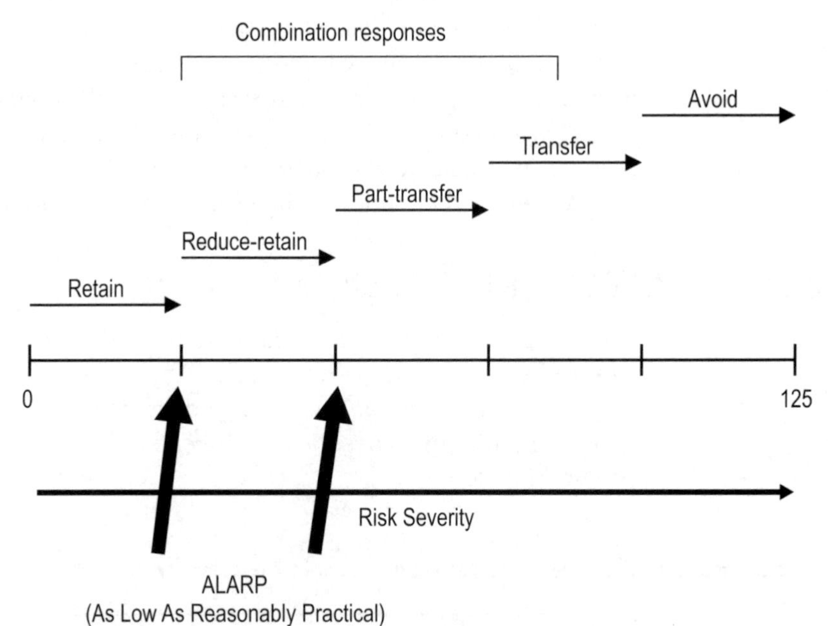

Strategic approaches to risk raise the issue of risk attitudes (or risk profiles) and how these can affect risk response decisions. This is further discussed in chapter 9. For now it is possible to proceed on the assumption that a typical project stakeholder organisation will tend to be averse to risk, and will thus adopt the ALARP principle. Whether it is risk averse or risk seeking, any organisation should arrange to monitor the risks it faces during the progress of a project, in order to place them firmly under continuous control.

8.4 RISK MONITORING AND CONTROL

An important, but often neglected part of systematic risk management is the ongoing monitoring and control of risks during the progress of a project.

Each project stakeholder should decide how to implement effective monitoring and control procedures for the risks it faces. Several issues must be considered, such as: determining which risks are to be monitored; assignment of responsibility; the type and frequency of monitoring required; reporting methods; the identification and treatment of new risks; and remedial or recovery planning and processes.

The first issue is most easily addressed, since all risks that the organisation has retained in any way should be monitored – at least to some extent.

Assigning responsibility for monitoring and controlling risks introduces an aspect of risk management that will be covered more fully in chapter 9. In any project stakeholder organisation, it is rare for one person (or one group of people) to bear responsibility for all the risk management activities of the organisation. As we have already noted in exploring risk response, the treatment of risks may need to be undertaken through financial management, quality management, sales management, safety management, and in many other areas of an organisation. This multi-management approach was illustrated in figure 6.1. However, it is also important to note that the people responsible for monitoring and controlling risks may be different to those involved earlier in identifying and analysing those same risks. Different risks may require different monitoring processes at different times by different people.

Table 8.1 introduced strategic management approaches at each level of risk severity, and this provides some guidance to assigning responsibility. The overarching principle is illustrated in figure. 8.2: the higher the risk severity, the higher should be the level of management assigned to deal with it. The type and frequency of monitoring and control procedures required will be influenced by the individual characteristics of each risk.

For negligible risks, minimum resources and staffing levels should be necessary. It is possible that monitoring and control processes for these risks could be combined with other routine activities in the organisation. Reporting will probably be by exception, i.e. only where observation indicates circumstances or situations arising beyond what could reasonably be expected to occur normally. Few data may be generated, and communication will be kept to a minimum.

At low levels of risk, monitoring and control activity may increase only slightly over that adopted for negligible risks. Monitoring is still

Figure 8.2 Risk severity / management responsibility levels

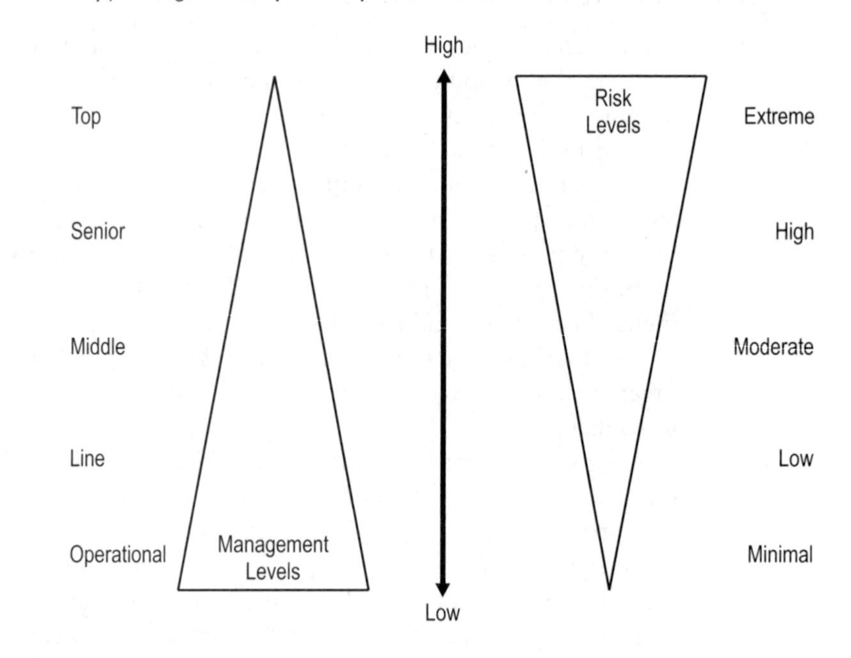

likely to be combined with other routine tasks, with the objective of flagging deviations from the norm. Communication will be on a more regular basis, with performance data relating to specific risks collected on a systematic basis.

For moderate level risks, monitoring and control should take on a more distinctive character. The frequency of monitoring would be clearly established in all the affected areas of the organisation, and formally incorporated into procedural policies and manuals. Staff involved should spend some, but not all, of their time in specific monitoring and control activity. Formal data recording, collection and reporting procedures are required, using planned communication channels.

For high severity risks it is advisable to appoint dedicated and trained staff to undertake monitoring and control activity on a regular, if not continuous, basis. Reporting should be to strictly defined standards and fast communication channels should be made available. Complex data may be generated and sophisticated data analysis techniques and data base access may be required. Emergency response and recovery plans and procedures should be in place.

With risks of extreme severity, the highest state of preparedness should exist in the organisation. Monitoring should be consistent and continuous. Senior management must be directly involved, and back-up personnel and other resources readily available as a standby. High level data collection and analysis procedures should be implemented in conjunction with secure data base access. Communication capacity should be extensive and flexible, allowing rapid and alternative means for urgent communication between any part of the organisation, with patch-in capability to external emergency services if necessary.

The processes of risk monitoring and control may reveal new risks for the project stakeholder organisation, requiring a return to the risk analysis and risk response stages of the risk management cycle. This repetition of the cycle may also be necessary for risks that have already been identified and treated, since it has already been noted that many risks are subject to changes in probability and impact over time.

In some respects, the monitoring and control process should add greater immediacy and urgency to the whole RMS, as risks are now closer in terms of the elapsed time of the project. On the other hand, there is greater knowledge about the project should be available and the uncertainty associated with many risks will be reduced as better information becomes available.

8.5 RISK DISASTER PLANNING AND RISK RECOVERY

Since even the best RMS cannot prevent every risk event from happening, project stakeholder organisations need to prepare and maintain plans to deal with the most severe risks they face.

Many issues are involved here, some of which include:

- assignment of responsibility for co-ordination and action,
- recruitment and training of emergency teams (and back-ups),
- defining key places, or routes for access, exit or congregation,
- defining key system control points,
- rehearsing other staff in emergency procedures,
- alerting of emergency services and specialists,
- provision of adequate communication facilities,
- supplies, equipment and spare parts logistics,
- public relations,
- statutory reporting requirements,
- preservation and collection of evidence,
- alternative arrangements for project delivery or completion.

The exact nature of each of these issues will be unique to particular types of risks – if not to specific risk events. Some relate more to physical risk environments, but recovery and disaster planning should not be limited to physical risks alone. For example, contemporary organisations in many fields have had to give considerable thought to the consequences of data loss arising from IT systems failure or vulnerability to attack.

More importantly, while there is some resemblance here to the issues of risk monitoring and control discussed earlier, it should be borne in mind that the recovery and disaster planning process is not the same as monitoring and control. Different people may be involved, at every level of the organisation. The reporting and decision-making processes may be different. Public relations requirements will certainly not be the same.

8.6 POST-PROJECT RISK EVALUATION AND RECORDING

Ignored even more often than the systematic monitoring and control of risks, the capturing of project risk knowledge is important for a stakeholder organisation. It must be able to learn from its experiences. Without some formal means of collecting information, processing it and placing it somehow into organisational memory, experiential risk knowledge will be left to reside in individual people in the organisation. Whilst this happens regardless of the existence of a RMS, failure to capture valuable information leaves the organisation vulnerable to the demise or departure of the personnel involved. Even if they remain active in an organisation, people often move from one project to the next, sometimes in rapid succession. Their memories will rapidly dim, especially where potentially negative aspects of previous project experiences are concerned.

A good RMS, therefore, should include the means to capture the risk 'stories' of the people involved in the project. Since members of the stakeholder organisation are also likely to have had interactions with people representing other stakeholders during the progress of the project, valuable risk learning from other perspectives may be acquired.

In most cases, post-project debriefing is an appropriate method of collecting risk information. Sometimes it may be appropriate to do this periodically throughout the project, perhaps when defined milestone stages have been achieved. While inter-organisational debriefing will be useful, the greatest value is likely to be derived from the deliberate encouragement and arrangement of intra-organisational project 'story telling' opportunities.

The information collection processes can be conducted informally, but it is important to formalise the analysis and archival recording of the collected material in order to maximise its usefulness to the organisation. The archival format may be incorporated into the formal project risk register, or dealt with as separate databases for each part of the organisation. We will examine this more fully in chapter 9.

8.7 COMMUNICATING RISK MESSAGES

The essential 'oil' for formal risk management is the degree of communication about risk that takes place throughout the project stakeholder organisation. High levels of communication are necessary if the RMS is to be effective. Risk messages and meanings must be clearly understood by everyone involved with risk management within the organisation. This is why the precision of risk statements is important during the risk identification stage. Equally important, however, is the plethora of intra- and inter-organisational communication necessary to deal with risk analysis processes and outcomes, risk response decisions, risk monitoring and control procedures, and the eventual capture of risk knowledge from projects and the people involved with them.

8.8 CHAPTER SUMMARY

This chapter has completed the presentation, from an operational perspective, of a broad framework for a project risk management system. It has been argued that project risks are really the risks of a particular stakeholder involved in a project, and that a single project-based RMS implemented by and on behalf of all stakeholders is not feasible, since different stakeholders will seek to fulfil different objectives, and will be engaged in different risk contexts, even for the same project. Each stakeholder in a project therefore needs to implement its own organisational RMS.

A formal RMS should incorporate clearly recognisable processes for identifying, analysing, responding to, monitoring and controlling the project risks faced by an organisation. It will assign responsibility at appropriate levels for exploring, evaluating and dealing with risks. An effective RMS will incorporate the planning of disaster and risk recovery procedures. It will also facilitate post-project capture of the risk knowledge people have gained from their project experiences. Good risk communication is essential throughout and beyond the organisation.

Many of the operational issues relating to formal risk management must necessarily be considered in the process of implementing, or building, a RMS within a project stakeholder organisation. This topic is addressed in the following chapter.

CHAPTER 9

BUILDING A RISK MANAGEMENT SYSTEM

9.1 INTRODUCTION

We have noted that, to an extent, we all (whether as individuals or as organisations) manage risk. The question that arises, therefore, is how well do we do it? This is really a matter of assessing the maturity of our risk management. That maturity in turn is a measure of how committed we are to managing our risks systematically; and of the system we have implemented, how it has been implemented, how it operates, how it is maintained and improved, and who has responsibility in all of this. Risk management is more a people issue than a mathematical conundrum. For many organisations, the risk management process focuses on managing the people who are dealing operationally with risks. Encouraging people to have an informed understanding about risk in general is one important aspect; getting them to adopt a formal approach to identifying and dealing with specific risks affecting their organisation is another, although it is not unknown for a key stakeholder (particularly a public sector client) to call for evidence of systematic risk management capability as part of its criteria for selecting potential participating stakeholders in its projects.

In this chapter we explore the implementation of formal risk management systems in project stakeholder organisations. Topics covered include: risk management maturity; organisational strategy and risk attitudes; clarifying objectives, tasks and commitments; creating the risk management system framework; assigning responsibility; system communication; trialling and evaluating techniques and procedures; learning from experience; and reviewing system effectiveness.

9.2 ORGANISATIONAL MATURITY IN RISK MANAGEMENT

The amount of system-building activity for risk management that takes place in a project stakeholder organisation is likely to correlate strongly with its level of risk management maturity. While not measurable in any precisely quantitative sense, an organisation's risk management maturity level should be discernible through careful observation of the ways in which it has established mechanisms to deal with project risk situations. Drawing upon the work of Hillson (2002), risk management maturity may be defined in four ascending grades, as illustrated in figure 9.1. The use of the present participle in each label is deliberate: it is intended to denote a sense of movement, rather than a static condition. Risk management should always be a dynamic activity.

Figure 9.1 Levels of organisational risk management maturity

Source: after Hillson (2002).

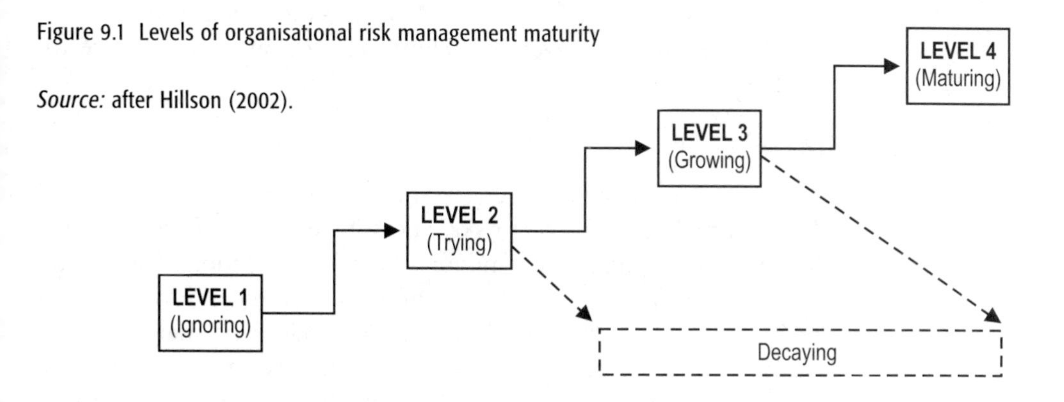

Level 1: Ignoring

Labelling an organisation as 'ignoring' in terms of its risk management maturity is perhaps not entirely fair, since every organisation does something about some risks. In a level 1 organisation, however, project risks are not dealt with systematically. Instead, risk treatment occurs sporadically, and usually on an ad hoc basis. More often than not the risk management will be reactive rather than proactive. By this we mean, for example, that project activity proceeds on the assumption that everything will occur as planned, and that any deviation from the plan will be 'dealt with on the day' since 'you can never tell beforehand what will happen'. In level 1 risk maturity organisations, the organisational level of project planning is itself likely to be rudimentary. 'Luck' may attach itself frequently to the organisation's descriptions of past project success, but failures are hardly ever mentioned.

Level 1 organisations have little awareness of needing to manage their risks, and thus possess no developed risk management culture. Nor, since 'ignorance is bliss', will they be fully aware of how they are responding to risks. They tend to respond only to external pressures. Matters such as occupational health and safety procedures, for example, may be regarded more as a necessary observance of regulations than as an integral part of an organisational risk management system. Often only the bare minimum for compliance is implemented. Many project risks are simply retained, without attempting any mitigating action.

In such organisations, any treatment of risk that does take place relies almost wholly on individual experience and expertise. Similarly, any risk learning from projects is rarely harvested organisationally, but is left to reside in individuals. Organisations with level 1 risk management maturity can survive, but their continuing survival itself becomes a matter of chance. From what has been written here, you may also deduce that a Level 1 organisation relies too much on clichés in its risk management. However, we have no evidence for that!

Level 2: Trying

In a level 2 organisation, there is at least someone who is aware of the need to manage project risks, and someone else (if not the same person) with authority to sanction management action. The awareness may have come through bitter experience, fortuitous or deliberate reading, conversations with colleagues or friends, newly appointed staff, or personal development activity such as attendance at seminars put on by the local chapter of a professional project management association. Alternatively, it may come about when the organisation is in an expansion mode and finds that it has to communicate aspects of its project decision-making internally or externally. It needs to summarise this in a way that will capture the implications for the organisation. Occasionally, the organisation may find itself bidding for project work where tenderers are required to submit risk management plans for the project. Such requirements are becoming more frequent in modern projects.

A fledgling or developing risk management culture should therefore be discernible at a reasonably high level within a level 2 organisation, but will not be visible throughout. Responsibility for establishing and operating a project-based risk management system (RMS) will have been assigned to someone, often in addition to his or her existing duties or on a short-term secondment basis. It is also possible that outside consultants will have been engaged to provide advice and assistance with implementing the RMS. More rarely, one or two staff members will have been encouraged to undertake some type of risk management training, although the type, level and quality of this training may be largely unknown.

Organisations at level 2 risk management maturity tend to set start-up limits for their newly implemented RMSs. This is done to contain costs (when the benefits are still largely unknown), and to create conditions that will permit subsequent comparison between RMS and non-RMS based projects, in order to assess whether or not the whole exercise has been worthwhile. Typical initial limits might include the adoption of formal RMSs only for projects above a certain value; for projects of a certain type (e.g. hospital construction projects, or banking ICT projects); or for projects carried out for a particular client.

While Level 2 organisations are seeking to systematise their risk management procedures, they will display a continuing reliance on reactive rather than proactive responses to risk. Risk knowledge capture is likely to be patchy and inconsistent, and any risk learning not yet widely disseminated within the organisation. Transfer may be the predominant risk response of Level 2 organisations, since they are still largely unfamiliar with mitigation techniques, and are largely unaware of the potential for exploiting opportunities in any of the risks they face.

Level 3: Growing

Within a level 3 organisation, risk management activity can be seen to be growing, but it is still almost exclusively project based. The majority of senior management is committed to the development of systematic risk management at all levels of project decision-making, even though this has not yet been fully achieved. A culture of risk management, and awareness of the need for it, permeates those parts of the organisation directly involved in projects. A separate risk management department may have been established. Specialist staff may have been recruited, and selected staff given the opportunity to undertake advanced risk management education and training, possibly at a postgraduate level.

Most risk management procedures in a level 3 organisation will have been standardised, and applied to all projects, although some unique or highly complex projects may test the capacity of the RMS. Proforma approaches to documentation will be evident. A cycle of improvement to some risk management procedures may be in train.

All projects will be subjected to risk debriefing at completion and a risk register, or risk knowledge database will exist. There may be evidence of fairly simple analysis of selected data, and risk information will be communicated at least across the operational parts of the organisation. Selected risk occurrences will be treated as learning experiences, as a means of internal training for staff and as the basis for procedural improvements.

Risk transfer responses will be subjected to increasing levels of cost–benefit analysis, and no risk will be retained without being subjected to at least a brainstorming attempt at reducing (mitigating) it.

On some projects, the potential for opportunity management will have been recognised, and limited exploitation may have occurred.

Level 4: Maturing

Risk management activity will be detectable throughout a level 4 organisation. While most visible in the operational, project-focused sections of the organisation, a strong culture of risk management will exist at all levels of decision-making. No project is allowed to proceed unless acceptable risk management procedures have been implemented for it, and no project lies beyond the organisation's capacity to assess its riskiness.

A separate risk management department may still exist, but it is possible that trained staff will now be found in all parts of the organisation, dealing with risk management issues in specific fields as an integral part of their other activities. Staff in the risk management department will play a more facilitating role, co-ordinating the post-project risk knowledge capture processes on every project, providing in-house training for the whole organisation; maintaining, and carrying out complex analyses of the risk knowledge database; conducting risk management performance audits; assisting other departments in establishing performance benchmarks; seeking to maintain a continuous improvement cycle for the risk management procedures; and communicating risk information throughout the organisation. The risk management department and the quality assurance department may have become seamlessly integrated. Review and maintenance of the risk management systems within the organisation will be regularly – and in some instances continuously – conducted, with responsibility assigned at the highest levels.

Analyses derived from the risk knowledge database will help the organisation to develop case studies for training purposes, and will be used to test different or innovative management techniques (not just risk management techniques). Strategically, the organisation will seek to exploit opportunities arising from such innovation, and from other risk situations. In part, the innovation will be used to increase the amount risk mitigation undertaken by the organisation, but a positive learning attitude will be taken towards any failures arising from the greater levels of risk retention. This in turn will contribute towards the organisation's ability to maintain a competitive advantage.

Decaying

Just as an organisation can mature in terms of its ability to manage risks systematically, so too can this capacity decay. In general, the higher the level of risk management maturity, the more gradual is the degradation of the system.

Decay is unlikely to occur slowly in an organisation at level 1 risk management maturity, since it is largely unaware of its maturity level anyway. Spectacular collapse is a more likely outcome, as luck is trusted once too often.

A level 2 organisation is likely to experience decay in its risk management activity for several reasons:

- Several projects have been completed without any clearly obvious benefit being derived from the application of the risk management system.
- The senior management champion for the RMS loses enthusiasm or leaves the organisation.
- Staff responsible for the RMS are swamped by other duties.
- The advice of outside consultants proves impractical.
- The post-project debriefings appear to yield little information of value.
- The RMS, or the risk knowledge database, proves too cumbersome or costly to operate and maintain.

In a level 3 organisation, decay in risk management maturity will be almost entirely due to lack of continuing commitment from senior management, or loss of key staff. The latter may also explain decay occurring in a level 4 organisation, but the problem in this case should only be temporary, given the organisation's risk management maturity and engagement with performance benchmarking.

9.3 ORGANISATIONAL RMS POLICY AND IMPLEMENTATION STRATEGY

Establishing a risk management system is a project in itself. There will be tasks, technologies, resources and organisation involved. Time will have to be allocated to implementing the RMS. Most importantly, there should be someone to champion the RMS-building project, and someone (not necessarily the same person) to manage it.

The RMS project may even experience all three project environments. Beyond the procurement phase of achieving an operational system, the RMS will have a functional life and maintenance needs. At some point it may be necessary to dismantle the existing system and replace it with a different one.

An important part of initiating a RMS in a project stakeholder organisation is the formulation of an organisational policy towards risk management and a strategy for implementing it. Ideally, this activity should start with a scan across the whole organisation, although in practice this is more often than not limited to the parts directly involved in projects. As noted earlier, an organisation's interest in risk management usually starts with a distinct project focus, and only later matures into RMS application to all of its activities.

An organisational scan addresses specific questions such as:

1. What (project) activities clearly require formal risk management?
2. How are decisions made about them?
3. What risk attitudes are evident?
4. What formal risk management is already in place?
5. How effective is it?
6. Is any informal risk management evident?
7. Where are the gaps in current risk management practice?
8. How could the gaps be filled?
9. Who should be involved in that?

The scan provides a risk management picture of the organisation in its current state. Almost certainly the picture will be fuzzy in places and lack a consistently clear focus. There will be evidence of activity resembling risk management in some parts of the organisation, but blank spaces in others.

Ideally, the scan should be conducted by bringing together appropriate representatives from within the organisation, and brainstorming the questions. Depending upon the size of the organisation, several meetings may be required. Within each meeting, large groups can be split into smaller sub-groups of up to five people. Each sub-group may be allocated a few questions (e.g. Q1, 2 and 3 to one group; Q4, 5 and 6 to another, etc.). Alternatively, each small group is required to address all questions. Small groups each dealing with a few questions may be best, since this encourages focus and discourages distraction, but each group should be aware of the full range of issues to be considered. Information about the meeting should be circulated beforehand, telling people about the purpose and intended topics to be addressed. Sub-groups should be reconvened into larger groups again after intervals of not more than about forty-five minutes, in order to report progress. All summary findings should be recorded in some way – by voice or video recording, whiteboard, butcher paper, or with secretarial help.

While representatives of senior management must attend these scan meetings, this does not mean that they should lead them. It may be more helpful to arrange for an independent risk management consultant to facilitate the activity and be responsible for preparing the agenda questions, recording findings and writing the subsequent scan report. If there is sufficient confidence and experience within the organisation, however, the facilitating role can be undertaken internally.

The scan should start by identifying project (or organisational) activities that participants believe should be subjected to formal risk management. It may be necessary to explain briefly the processes of formal risk management before exploring this question. Note that the question does not ask participants to identify *every* project activity the organisation engages in; but only those where formal risk management is desirable. This is deliberate, with the intention of avoiding a situation where people could become overwhelmed with too much detail. Some activities may be overlooked, of course – hence the blank spaces referred to above. The missing parts will eventually be filled in as a culture of risk management develops within the organisation.

The second question clearly begins to associate decision-making with risk management. The answers produced by the scan groups should point to any mismatches occurring in decision-making levels within the organisation, and will help to identify other staff who might usefully become involved in subsequent formal risk management implementation processes.

The third question of the organisational scan introduces the concept of risk attitudes. Classically, in decision science and behavioural psychology, people are described as being either risk averse, risk seeking, or risk neutral. In reality, the latter attitude is most unlikely. People either like to buy a lottery ticket or they don't: they are keen to engage in extreme sports or they are not. You will hardly ever meet anyone who is truly indifferent in his or her attitude towards risk, regardless of the context.

Risk attitudes are revealed in decision-making – thus providing an additional purpose for question 2. A person might intimate that he or she is risk averse, but in real life make decisions which actually reveal a tendency towards risk seeking.

Attitudes towards risk are also complicated by other factors. A person's risk profile is not necessarily consistent over the whole range of his or her decision-making. Some decisions will be dealt with in a risk averse manner; others will receive a risk-seeking response. The risk profile is not necessarily consistent over repeated or similar decision circumstances, nor over time. Many decision-makers are not consciously aware of their risk attitude for most of the time.

Education, training, culture and experience all combine to shape our risk attitudes. Most adults are capable of separating professional risk attitudes from personal risk attitudes where the distinction is necessary and the attitudes differ. In turn, project organisations that tend to be risk averse rely on this capacity in the staff they employ at higher decision-making levels.

In effect therefore, the third question is something of a Pandora's box, since it may lead to a labyrinth of psychological and behavioural connections that cannot be unravelled in the limited environment of an organisational scan. The purpose of the question is to expose at least some of the influences on organisational decision-making, as these will also influence the development of a policy towards risk management.

Providing participants have a reasonable grasp of formal risk management processes, the fourth question should be fairly straightforward. In most cases, the answers will reveal patchiness across the organisation. Some aspects of risk management will be evident – albeit under different labels – in some existing project or organisational activities such as occupational health and safety management, quality management, value management, financial management and purchasing. Others will be clearly absent.

Ascertaining the effectiveness of any existing risk management practices (question 5) is almost certain to produce highly subjective and qualitative responses, since the organisation is likely to have few, if any, quantitative measures in place. This question is aimed at getting participants to describe their confidence in existing organisational processes, but is posed in a rather indirect manner to encourage people to give open responses.

The sixth question could also be rephrased to ask what intuitive or non-standard procedures are used, when making or implementing project decisions, to ensure that satisfactory outcomes are achieved and objectives fulfilled. The purpose of the question is to identify effective procedures that could later be incorporated into the formal RMS, and to weed out weaknesses.

The last three questions may be best left to a post-scan session, since some expertise in assessing formal and informal risk management systems is required. It may be easier for a small senior management team from the organisation, perhaps with specialist assistance, to analyse the information gained from the earlier scan questions and, with their knowledge of the structure of the project stakeholder organisation, identify inadequacies in its current risk management practices. However, if the participants in the scanning sessions are maintaining their enthusiasm for this forensic work, they should be encouraged to continue and suggest how procedures could be improved.

The outcome of an organisational scan provides a richer picture of the state of risk management practice within the organisation and guides the directions in which it can be formalised and improved.

The project stakeholder is then in a position to formulate (and document) a coherent risk management policy. Documentation of risk management policies and procedures is an important aspect of corporate governance, since it facilitates communication, accountability and subsequent performance review.

The policy document should be specific rather than general. It should state why the organisation wishes to adopt a more formal approach to managing risk, and indicate if the RMS is to be targeted at particular projects, to all projects, or even across all the activities of the organisation. Where possible, risk management responsibilities should be defined and assigned, using an organisational structure diagram if necessary.

A RMS implementation strategy should include measures to motivate staff. Personnel who will be involved in using the RMS should be represented in the implementation team. Too often, the introduction of a formal RMS is seen as simply another imposition that must be added to an existing workload. If necessary, of course, workloads and remuneration should be adjusted to reflect additional duties and responsibilities but, more importantly, employees should be shown how engaging in effective risk management not only improves their skills but could also contribute towards greater job security by reducing the organisation's business vulnerability (or enhancing its performance).

The RMS implementation strategy should also consider the nature and extent of risk management education and training required, and how risk management knowledge is to be diffused throughout the organisation. The objective of the strategy should be to grow a risk management culture within the organisation, by raising awareness of project risks, and by enabling staff to develop skills and confidence in managing them in a systematic and transparent manner.

9.4 CLARIFYING OBJECTIVES, TASKS AND COMMITMENTS

As with any project, the implementation of a RMS within an organisation should be carefully planned. Objectives must be clarified, tasks scheduled, commitments confirmed, and resources obtained and organised.

Ideally, the procurement objectives should state what level of RMS is to be implemented, at what budgeted cost, and by what intended operational date. Functional objectives should indicate what the RMS is intended to do (e.g. facilitate the identification, analysis, treatment, monitoring and control of project risks faced by the organisation; assist in the necessary documentation of the techniques employed, the decisions made and the outcomes achieved in the risk management process; and become a repository of organisational knowledge about project risks). Strategic objectives will describe what the outcomes of the RMS are expected to achieve for the organisation.

Scheduling the tasks for implementing a RMS largely depends upon how the system is intended to operate within the organisation. This issue was mentioned earlier, and it is useful to expand upon it here. Essentially, there are three ways of operating a RMS:

- as a single, centrally based system to deal with all the activities of the organisation;
- as dual, centrally based systems: one to deal with project activities; the other to deal with internal organisation maintenance activities;
- as multiple, separate systems for each project; plus a single system for internal organisation maintenance.

A single organisational RMS may be achievable only for an organisation with Level 4 risk management maturity, where a seamless organisational approach is sought. The dual system will be found in organisations with level 3 risk maturity; while a level 2 organisation is likely to be trialling multiple RMSs – at least for selected projects – and may not yet have decided to implement an additional system for its internal activities. Initially, therefore, risk management system building is most often aimed at individual projects.

The tasks of RMS building will involve staff at all levels in the organisation, and implementation will require activities involving:

- system design (How will the RMS operate? What sort of framework is desirable? How should it be implemented?)
- assigning responsibility (Who will be involved?)
- deciding the level and nature of communication channels (How is risk reporting to occur? To whom? How often?)
- trialling techniques (What risk identification and assessment techniques will be used?)
- evaluating the implications of risk response options (How will each type of response impact upon the organisation?)
- evaluating the implications of risk monitoring and control activities (How will these affect the organisation?)
- designing and establishing risk registers (How should risk knowledge be captured?)
- establishing criteria and procedures for reviewing risk management effectiveness (What are the RMS performance criteria? How should they be assessed? By whom? How often?).

None of these activities can be undertaken successfully without the committed support and involvement of senior management. Without adequate support, the necessary risk management culture will never become completely ingrained throughout the organisation. Nor is support sufficient without appropriate involvement, since the association of risk with decision-making means that decision-makers at all levels in the organisation must participate (and be seen to participate) in the risk management process.

9.5 CREATING A PROJECT RISK MANAGEMENT FRAMEWORK

The system framework for implementing risk management can be as simple, or as sophisticated, as the requirements of the organisation warrant.

At a fairly basic level, computer spreadsheets can be used to deal with individual projects. Typically, a spreadsheet format for the RMS might resemble that shown in table 9.1 (A and B). The format can be developed as a template for the organisation and its projects. The left-hand part of the spreadsheet records the nature and characteristics of identified risks. The centre section is used for assessing these risks; the right-hand columns reflect decisions made about their treatment, monitoring and control.

Depending upon the level of risk management implemented, the first few columns of the RMS spreadsheet are used to list project objectives and the tasks, technologies, resources and organisation necessary to achieve them. This information may be gathered either from existing project documents, or derived from the collective ideas of project team members.

A key column is allocated to recording every risk associated with the objectives and project elements and sub-elements. These should be identified through project team brainstorming and other techniques. The identified risks can be immediately entered into separate spreadsheet cells during the brainstorming activity, but subsequently each entry should be edited to create precise risk statements couched in the terms suggested in chapter 6.

The column used to record the type of each identified risk is a useful way of discovering if the project is significantly vulnerable to particular categories of risk. It can also help in the subsequent exploration of alternative options for treating these risks. This column does not need to be completed during the risk identification sessions.

In the assessment section of the spreadsheet, three columns are included to allow subjective scoring the probability, impact and exposure duration dimensions for each risk, using simple 5-point scales. Ideally, the scoring should be completed during the risk identification session, but it could be carried out during a separate subsequent session, especially if considerable editing work is needed to produce the necessary precision in the risk statements. A fourth column uses simple cell formula entries to calculate a severity score for each risk. At this point, it is wise to save the current version of the project RMS computer spreadsheet, and then re-save it as a later version in the project file sequence.

The sorting capacity of most commercially available spreadsheet software applications permits a simple re-ordering of the scored risks, ranking them from highest to lowest severity score. The organisation can now identify the most severe risks it faces on the project, since these will be clustered at the head of the spreadsheet. Alternatively, the most risky objectives, tasks and commitments for the organisation on that project can be highlighted. It is then possible to revisit these if necessary, armed with greater knowledge, thus providing the opportunity to rethink, and possibly improve the organisation's performance. Although it has been omitted from table 9.1B in the interests of saving space, an additional column could be inserted at this point to record the severity of each risk in terms of the organisation's strategic risk management descriptors (table 8.1).

The system framework should include columns to record any procedures or controls already in place for the risks that have been identified. Typical entries here might include contract clause numbers or specification references. In the spreadsheet template, this column marks the beginning of the treatment section. Much of the cell entry data in these columns will be in text form.

Knowing the extent of the risk treatment procedures already in place allows the risk management team to consider potential treatment gaps for each of the risks that have been identified; and then to suggest alternative or additional methods of dealing with them. Another useful column (also omitted from table 9.1B) could be inserted after this point to record the type of risk treatment option suggested (i.e. avoid, transfer, reduce and retain, retain, or combinations of these). This allows immediate comparison with the organisation's preferred treatment strategy for risks of particular types and severity.

Where it is appropriate, any identified risk can now be reassessed, using the subjective scoring approach as before. This is a crude but acceptable way of gauging the potential effectiveness of the proposed risk treatment option. By saving successive versions of the template RMS computer file, it is always possible to revert to earlier choices.

The remaining columns of the RMS framework template are used to record the actual treatment decision for each risk, how monitoring will be carried out, and how responsibility is to be assigned. It may even be possible to include information about the costs of risk treatment. This will permit subsequent evaluation of the benefit-to-cost

Table 9.1A Simplified project RMS format

PROJECT:								
Risk ref.	Objective context	Element context	Sub-element context	Identified risk	Likelihood score (1–5)	Impact score (1–5)	Duration score (1–5)	Severity score (ex 125)
01								
02								
03								
04								
05								
06								
07								
08								
09								
10								
...								

Key:

LIKELIHOOD	IMPACT	DURATION
1. Rare	1. Insignificant	1. Short term
2. Unlikely	2. Minor	2. Medium-short term
3. Possible	3. Moderate	3. Medium term
4. Likely	4. Major	4. Medium-long term
5. Almost certain	5. Catastrophic	5. Long term

Table 9.1B Simplified project RMS format

PROJECT:											
Risk ref.	Severity score (ex 125)	Ranking (1 = most severe)	Existing risk treatment	Missing treatment	Proposed treatment score	Revised severity	Response decision	Monitoring	Responsibility	Benefit $	Cost $
01											
02											
03											
04											
05											
06											
07											
08											
09											
10											
...											

performance of the RMS, although consistent accuracy should not be expected in such assessment.

The simple spreadsheet approach described here will probably satisfy the early risk management system requirements of an organisation with level 2 risk management maturity. With adaptation, it can be made to serve the needs of organisations with greater risk management maturity. It is possible, for example, to use cell entries to reference links to other spreadsheets, schedules and documents where material relevant to each risk is dealt with in greater detail. Organisations with level 4 risk management maturity are likely to have developed more sophisticated risk management system frameworks, and may well be using intra- and extra-net computer information technology networks to process and communicate project risk information within and, where necessary, beyond the organisation.

9.6 ASSIGNING RMS RESPONSIBILITY

The need for the support and involvement of senior management – both for building and operating a RMS – was argued earlier in this chapter on the grounds of the association of risk with decision-making. Additional arguments lend cogent support to this need.

An appropriate culture of risk management is desirable throughout the organisation. Such a culture recognises the importance of dealing effectively with potential threats to the organisation through their impacts upon the projects with which it is involved. While the threats may be negative, however, the management process for dealing with them needs to be positive in terms of a forward-looking approach which instils confidence in the way the interests of the organisation are being safeguarded.

Instilling and growing such a culture is a top-down leadership task and is the responsibility of senior management. The active involvement of staff at this level, willing to drive the system building project for the RMS and prepared to champion its adoption and development, provides the best chance of successfully implementing an effective RMS within an organisation.

The appointment of a risk manager does not automatically resolve all risk management responsibility issues within an organisation, despite the fact that advertisements in the 'situations vacant' columns of many newspapers indicate that surprisingly large companies appear to make this assumption. While such an appointment can contribute substantially to the effectiveness of the RMS – whether at the implementation stage or for system operational purposes – the risk manager does not thereby become responsible for all the decision-making occurring within the organisation. Nor could he or she possibly undertake all the necessary risk management activities alone. Risk management is never the exclusive preserve of one person.

As the organisation's risk management maturity increases, it will become evident that the RMS operating team will be substantially different from the RMS building team. Ideally, the people necessary for building a RMS will be found within the organisation itself. Members of the system building team should be familiar with the structure of the organisation and its activities. They should be able to communicate effectively across any departmental or discipline boundaries existing within the organisation. Initially it may be best to start with a small team and allow this to grow organically with the development of the RMS itself. Help may be needed at the outset.

This raises the issue of using external consultants in risk management. While risk management specialists can make a valuable contribution to the implementation of a risk management system (and advising on its application to specific projects), it is important to clarify their role and the extent of their involvement. Remember that, regardless of the number of consultants you appoint, as a project stakeholder these are your projects – not theirs – and therefore your risks, not their risks!

Specialist risk management consultants can assist in RMS building by facilitating the workshops needed to carry out backward and forward organisational scans and undertake SWOT analyses; and to assist with training staff in the use of risk management techniques. The help of consultants may also be necessary for tricky or unusual projects, particularly during the risk identification and assessment stages for projects where the existing risk management expertise in the organisation may be inadequate.

Personal knowledge or word of mouth may be the only means for an organisation to judge the suitability of potential risk management consultants, since currently there is no professional accrediting body for this discipline and micro-specialisms (e.g. safety risk management or financial risk management) proliferate. A specialist in one area is not necessarily competent in others. Careful selection of an appropriate consultant is therefore necessary, and it is important to specify the nature of the services sought and the duration of the appointment. Evidence of qualifications and previous experience should be examined, and references obtained from the consultant's recent clients.

9.7 RISK MANAGEMENT COMMUNICATION

Effective communication is a critically important ingredient in risk management, since risks can only be managed if they are known and understood, and knowledge and understanding can only occur if the participants in the risk management process are able to find common meaning about the nature and extent of each of the risks that they must deal with. Risk communication must be effective intra-organisationally (i.e. within the project stakeholder organisation). More often than not, it also needs to be inter-organisationally adequate (i.e. between project stakeholder organisations).

The potential extent of risk communication required can be seen on projects undertaken under the Private Finance Initiative (PFI) in the United Kingdom. This scheme has been used extensively since 1992 to secure the delivery by private sector organisations of public projects and services on the grounds that such delivery can be more efficiently and economically managed by the private sector. It is part of a worldwide increase in the involvement of the private sector in the development and financing of public facilities and services. Procurement arrangements, under the general label of 'Public Private Partnerships' (PPPs), have been developed to encourage the public and private sectors to work together to share the risks and rewards associated with such procurement activities. These PPPs range from the simple contracting-out of public services to the involvement of the private sector in the financing, design, construction, operation, maintenance and, in many cases, concessional ownership of major facilities such as roads, tunnels, water supply infrastructure, hospitals, clinics and schools. The effectiveness of PFI has yet to be fully demonstrated over the complete life span of an acceptable sample of similar projects, and some doubts have been raised about many of the advantages claimed for it. In the United Kingdom however, it is recognised as a means of avoiding centrally imposed limitations on the extent of public sector local authority borrowing, since the financing of projects becomes the responsibility of the private sector partners and can thus be treated as off-balance sheet items by the public sector clients.

Under a typical PFI arrangement, a Special Purpose Vehicle (SPV) is created by the private sector collaborating partner organisations to deliver the required project (Akintoye et al., 2001). A typical arrangement is shown in figure 9.2. The SPV is a legal entity formed with the single purpose of delivering the project, service or facility. Where this involves an eventual reversion of ownership to the public sector after a fixed period, for example in situations such as build-own-operate-transfer (BOOT) projects, the SPV has a finite life-span.

The process of implementing a PFI scheme, from the perspective of the public sector client, is illustrated as a flow diagram in figure 9.3. As a measure of public accountability, before a PFI project is allowed to proceed beyond the development stage in the United

Figure 9.2. Typical organisation structure in public/private partnership projects (UK Private Finance Initiative model)

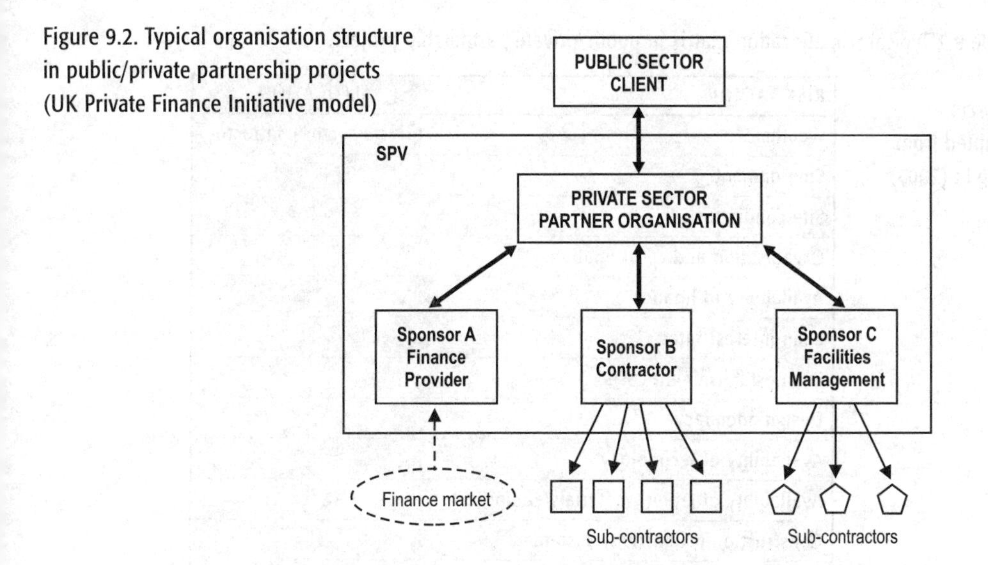

Figure 9.3 Typical pre-contract development process in public/private partnership projects

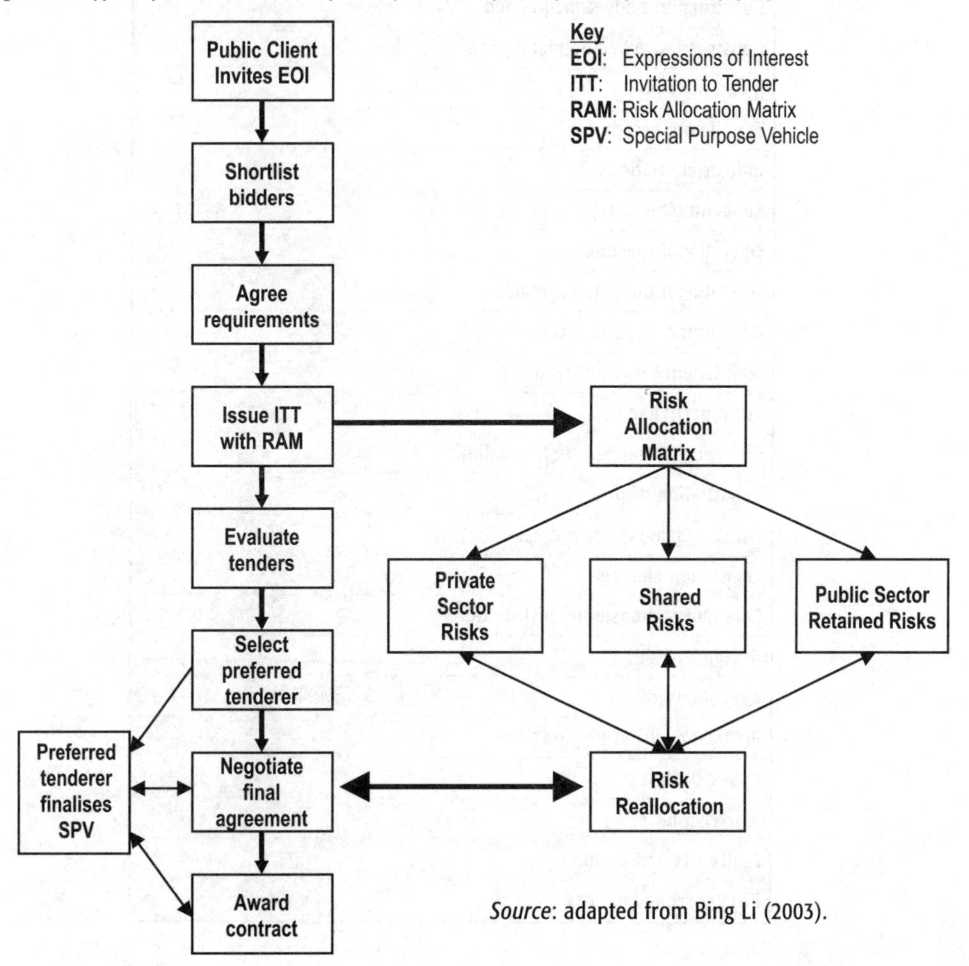

Source: adapted from Bing Li (2003).

Table 9.2 Typical risk allocation matrix in public/private partnership projects

Source: adapted from Bing Li (2003).

RISK FACTOR	ALLOCATION
Weather	Private sector contractor
Environment	"
Site conditions	"
Organisation and co-ordination	"
Availability of finance	"
Loan interest rates	"
Loan establishment costs	"
Design adequacy	"
Availability of technology	"
Availability of labour/materials/equipment	"
Construction productivity sustained	"
Quality assurance	"
Construction budget maintained	"
Construction schedule maintained	"
Inflation	"
Tax climate acceptable	"
Industrial relations	"
Subcontractor default	"
Operational revenues	"
Operational budgets maintained	"
Operational productivity sustained	"
Maintenance quality sustained	"
Third party liability	"
Currency exchange rate fluctuations	"
Repatriation of profits	"
Level of demand for project	Shared
Legislation change	"
Role and responsibility distribution	"
Design changes	"
Availability of site	Public sector client
Political decision-making	"
Public protest	"
War/civil unrest	"
Approvals and permits	"
Project scope changes	"

Kingdom, the client must demonstrate to central government authorities that the PFI proposal compares favourably with alternative procurement methods including fully public funded delivery.

The risk allocation matrix referred to in figure 9.3, will include a number of different risk situations, indicating how these are to be dealt with on an allocation basis between the parties but omitting any treatment detail. Typical risks allocated in this way are listed in table 9.2.

Each of the project stakeholder organisations (and there will usually be many others) must ensure that it is fully informed about the risks it faces on the project. The public sector client must ensure transparency of risk information in the documents issued to bidders and partner organisations. The participating collaborators in the SPV must ensure that they share a common understanding of the risks each of them has been allocated.

Note that the reference to 'shared risks' in this PFI example does not mean that the process of risk management for these risks will be undertaken jointly by the public sector client and the project SPV. The principle enunciated in chapter 8, that each stakeholder is responsible for managing its own risks, still applies. Each participant will manage its part of each shared risk, although the stakeholders may confer over the assessment and treatment options for these risks.

The PFI example provides a broad indication of the potential extent of intra- and inter-organisational risk communication that might be necessary on a project. It is difficult to estimate accurately how many individual risk-related messages might be involved. Analysis of 'extra-net' examples of information management systems in the United Kingdom, for more than 150 projects, has revealed that the average project can involve storage of upwards of 15 000 pieces of information in the extra-net, including common information documents made available to all participating project stakeholders as well as inter-stakeholder information transactions. Of the 15 000 items stored in the project extra-net, about 50 per cent are claimed to represent graphic information (project drawings) and the rest include single page and multi-page documents. A significant proportion of the extra-net material will be concerned with project decisions and thus have risk implications. If only 10 per cent of the text-based system entries contain explicit risk-related messages, then at least 750 pieces of project information could be directly connected with risk issues, without reckoning on those among the remaining 95 per cent of all entry items that might have indirect risk implications. This suggests that the 'iceberg' simile is applicable to current levels of project risk communication: far more is likely to be occurring implicitly than explicitly. Any RMS building activity must therefore involve consideration of how effective communication can be achieved, at least within the project stakeholder organisation itself and preferably beyond it.

In terms of risk communication, therefore, the RMS building process should ensure that sufficient precision and reliability are incorporated into all of the media-related aspects of the RMS, including the system framework itself and all supporting identification, assessment, treatment, control, monitoring and register documents. Similar precision and consistency will also have to be sought in documents accessed beyond the organisation, including plans, schedules, contracts, subcontracts, warranties, estimates, tenders and such like.

It is vital that the risk-related implications embedded in all these documents be clearly perceived and understood as risks, i.e. that they properly reflect the threats facing the project stakeholder.

9.8 TRIALLING TECHNIQUES

An organisation lacking any formal system of risk management is likely to be unfamiliar with appropriate techniques, especially in the areas of risk identification and risk analysis where intuition and personal experience were previously relied upon. Thus there is a need to develop the techniques required to make the RMS effective in operation. The task of developing precise risk statements can be daunting to anyone who has never done it before, and even brainstorming might require some practice among staff unaccustomed to this approach. Practice in dealing with group dynamics may be necessary. It is also not unusual to encounter confusion about particular concepts of risk, such as uncertainty and probability, among the staff of an organisation, so the trialling of techniques may need to be accompanied by education and training.

Different techniques should be trialled during the RMS building process, using the assistance of specialist risk management consultants if necessary. Ideally, staff should be encouraged to experiment with as many different techniques as possible (e.g. the methods of risk identification discussed in chapter 6), and feedback obtained about the perceived effectiveness of each. An important aspect of this work is the alignment of any subjective rating scales intended for risk assessment purposes, so that they match organisational values, followed by calibration to ensure that they are applied consistently across the organisation.

Project simulations and case studies are good vehicles for trialling risk management techniques, particularly where tricky or critical project situations can be tested. It is advisable to adopt an incremental approach to these trials, starting with techniques that are sufficiently simple and then gradually building up to more sophisticated and complex methods. Early mistakes on the part of participants must be forgiven, and enthusiasm rewarded. Errors should prompt critical review of the technique in question, rather than blame for the users.

The process of trialling risk management techniques should be carefully planned and properly resourced when building a RMS, since starting a system with inadequate techniques may damage it to the extent that recovery is almost impossible. Even where appropriate techniques have been implemented, wise organisations with good risk management maturity will continue to research and develop this aspect of their systems.

9.9 EVALUATING RESPONSE OPTIONS

The four risk response options described in chapter 8 (avoid, transfer, reduce and retain, retain) sound deceptively simple, and even knowing that combinations of some of these are possible does not appear to add undue complication to the task of deciding how risks should be treated.

In practice, decisions about risk response options are rarely straightforward since almost every response will entail ramifications for the organisation. Avoiding, transferring or mitigating some risks may incur additional and different risks which in turn must be identified and managed.

It is important therefore to evaluate response options to risks carefully in terms of their flow-on effects. As with other risk management techniques, risk response options can be explored in a simulation or case study environment during the RMS implementation stage. Senior management input is essential during this process, since staff at lower levels may be unsure about what constitutes an acceptable or unacceptable risk.

9.10 EVALUATING MONITORING AND CONTROL PROCEDURES

The procedures required for monitoring and controlling project risks will be determined by the nature of individual projects and the specific risks they impose upon the stakeholder organisation. Nevertheless, some thought should be given at the RMS system building stage to the general implications of this task. Most of the system building team will have project experience with the organisation, and will be familiar with at least some of risks involved in those projects and the measures that were used to deal with them. In many instances these would have comprised management procedures from other parts of the organisation, such as financial control, quality assurance and occupational health and safety management. The team should therefore be in a practical position to consider how the implementation of a formal RMS in the organisation might affect procedures already in place, and to suggest any changes or additional resources necessary to improve their effectiveness from a risk management perspective.

Clearly, the RMS should avoid unnecessary duplication of existing management control processes, but not at the expense of failing to properly facilitate adequate management of the project risks of the organisation. A recent example, albeit not from a project management context, emphasises the need for effective control. Early in 2004, a major Australian bank conceded losses of AU$360 million in foreign currency option trading. A subsequent independent auditor's report blamed, among other things, a complacent and arrogant corporate culture within the bank, management deafness and indifference to earlier warning signals, misleading reports communicated to senior management, and the by-passing and manipulation of the bank's risk management systems. The lack of banking experience among the non-executive board members of the bank was noted. Apart from affecting the bank's share price, this damning report sent shivers through the whole Australian banking industry. Risk management which is superficial, or which can be ignored or manipulated, is itself at risk of being ineffective through failure to adopt appropriate monitoring and control processes. Not unexpectedly, the first response to the bank's loss was the resignation of its chairman and the chief executive officer. The new chairman's subsequent reaction to the auditor's report was to dismiss several managers and staff and reassign the duties of some directors. Instead of attempting to placate shareholders by the traditional 'blame game', the bank might have served them better by providing clear explanations of the bank's proposals to improve its risk management systems.

9.11 ESTABLISHING RISK REGISTERS

The more care and attention that is devoted to the design and establishment of the risk register, the more valuable this component of the RMS will be to the stakeholder organisation. Companies worldwide are now beginning to appreciate the benefits that knowledge management can bring, and a key part of this is the means of capturing the knowledge that is developed and resides in the organisation. For the most part, risk knowledge lives within individual members of the organisation and, while there may be ethical issues of ownership to resolve, the organisation should make every effort to harvest and exploit the available risk learning.

For some project stakeholder organisations, a risk register comprising an updated version of the spreadsheet illustrated in table 9.1 might be sufficient, especially where relatively simple projects with few risks are involved. As an organisation's risk management capability matures however, it is likely to need a more extensive means of recording project risk outcomes and experiences.

In designing a risk register, several aspects must be considered:

- What are the objectives?
- Who will use the register?
- How should the register be structured?
- What data should be collected?
- Who will collect it?
- What are the data sources?
- How should the data be collected?
- How should the data be analysed?
- How will processed data be used?
- Are there issues of confidentiality?
- Will the information have a 'shelf life'?

Objectives

The objectives of a risk register will normally be associated with:

1. creating a repository of useful risk-related information gained from project experiences;
2. making this information available for beneficial use on future projects;
3. facilitating the use of risk information to guide change in the organisation;
4. ensuring that valuable organisational knowledge is not lost through staff turnover;
5. providing an accessible and auditable database of organisational risk management practice.

The first two objectives are clearly important in terms of the focus of an organisation whose main activity involves a stakeholder role in projects. The third objective recognises that project experiences are not necessarily limited to their potential effect on future projects, but can also influence change in other areas of the organisation. An instance of financial risk occurring on a particular project, for example, might result in changes to the financial management procedures for the whole organisation.

It is surprising how few organisations take steps to preserve valuable organisational knowledge and prevent it from being lost when key staff leave. A risk register is a good counter to this risk.

Similarly, organisations should recognise that a properly maintained risk register is capable of providing powerful evidentiary support in the event of inquiries held to investigate project incidents. An inquiry is likely to show a more sympathetic attitude towards an organisation able to demonstrate both proactive and reactive competence in dealing with risks.

Users

The primary users of a risk register will be the decision-makers in the organisation, since it has already been established that project risks arise out of project decisions.

Structure

The structure and format of a risk register are critical to its effectiveness. Ideally, the register should be a 'live' repository of information; a resource that grows with the organisation and is flexible rather than rigid. It should be capable of accommodating a wide range of uses and users, be searchable, and preferably be selectively interactive rather than entirely passive.

This suggests that an electronic platform for the register is likely to be more efficient than a printed hard-copy manual, at least for the majority of intended uses. A computer-based intra-net information management approach is probably ideal, since this allows control over access within the organisation, has data storage capacity and permits interactive operation by selected users and passive access by others. Security and flexibility of access can be assured.

The risk register can be structured in various ways according to the perceived needs of the organisation. For example, it could be formatted primarily in terms of project type. This might suit an organisation which tends to engage repeatedly in several different types of project, but would be less useful for one that restricts its activities to a single type (or very few types) of project.

Another approach might be to adopt risk classification types as the primary denominator for the register. Thus separate sections of the register would deal with different risk categories such as weather risks, social risks, political risks, cultural risks, legal risks, technical risks, etc., according to the risk classification system adopted by the organisation. Risk classification was discussed in chapter 2, and this method of structuring the risk register would suit an organisation involved in different types of project but tending to experience the same types of risk on each project. This approach would also be appropriate for a project stakeholder organisation operating wholly within an industry with industry-specific risk types.

A third alternative would be to adopt project elements and sub-elements as primary structural determinants for the risk register. The task, technology, resource and organisation elements of all projects were discussed in chapter 3. In practice, this approach would be most useful for an organisation which is generally engaged in a limited number of activities that are similar for all its projects.

Finally, the structure of the risk register could be based primarily upon the project environment (e.g. procurement, operation or disposal as discussed in chapter 3) or upon the phases within any of these environments. In terms of the latter approach, for example, divisions in the register might reflect the risks associated with the design, testing, tendering, installation and commissioning stages of a project.

A key to deciding upon the basic structure of a risk register is to ascertain the general approach to risk identification used by the organisation in its RMS. If the RMS tends to favour a risk identification technique based upon one of the approaches noted above, then this should influence the format adopted for the risk register.

It should also be noted that none of these approaches envisages the incorporation of the organisational risk register and the RMS for each project. While it is useful for each project RMS to be prepared and used with eventual archival purposes in mind, it would create an unwieldy tool to attempt to combine the project RMS with the organisational risk register.

Data

In chapter 8 we referred to the post-project risk evaluation and recording stage of an organisational RMS as 'getting the risk stories'. While this implies an anecdotal quality for the information involved, it certainly does not mean that fictional accounts are acceptable. The data needed for the risk register of a project stakeholder organisation's RMS may be qualitative or quantitative in nature, but should all be based upon factual measurement, observation or objective opinion.

For the most part, data needed for the risk register will relate to any or all of the components of particular risks: the circumstances (including background details of relevance) and nature of the risk event and its occurrence, consequences and period of exposure. Information about the effectiveness of any pre-planned mitigating procedures should be included, as well as details of unplanned actions that became necessary. It is obviously critical to include full accounts of any disaster events and the recovery measures taken. Cause and effect observations are extremely useful.

Most of the data are likely to relate to risks identified and treated in the project RMS, but care should be taken to gather information about new and additional risks arising during the project activities.

The 'richer' the data (i.e. beyond mere facts and figures) collected, the more valuable the information will be for the organisation.

Data collection

How and when risk data should be collected, and who should be involved in this task, are clearly matters for each project stakeholder organisation to determine for itself.

Generally, a post-project debriefing process (or project post-mortem) is a convenient way of gathering risk stories. However, it is not necessary to wait until a project is finished before undertaking this task, nor should it be assumed that a short, single unbroken process will be sufficient. Nowadays, many projects have a post-project meeting for all key stakeholders. While useful risk-related information will almost certainly be gained from this, an organisation keen to improve its risk management performance should arrange additional in-house occasions to collect data. These may also be undertaken during the currency of the project, perhaps at significant milestones. Besides group meetings, individual interviews with project staff may be necessary.

The people responsible for collecting data should be prepared to do this actively rather than passively. In other words, it is better to go out and seek information and not to assume that others will bring it to you. For disaster and disaster recovery situations, of course, an active data collecting and recording intervention is essential.

Collecting project risk data requires people who can listen to others' stories with patience and accuracy, and who have an ability to organise information coherently. Some of these are acquired skills, so a prudent organisation will arrange for suitable training for its risk management staff where necessary. Sound recording can be used during group meetings and individual interviews, but not all risk data will be verbally based and opportunities (and the capacity) to use other media for direct data collection, such as video and digital photography, should not be overlooked. Nor should secondary sources beyond the organisation and its projects be ignored: newspaper and journal articles, statistical and official publications, television documentaries and radio broadcasts can all yield useful information about risk.

Analysis and use

Besides the obvious 'what to do/not to do on the next project' value of the risk register, there are other potential areas of analysis and use for this resource.

Quantitative analysis, however limited this may be, should be possible for at least some of the risks of similar type encountered on several projects. Where the projects are sequential it may also be practicable to assess the effectiveness of any treatment procedures implemented over the intervening period. Time series analysis might be attempted where this is appropriate – for example, for some economic risks.

An important use of the risk register lies in the development of risk case studies, scenarios and disaster recovery plans. Because the risk information is so cogent to the organisation itself, it is capable of providing realistic material that can be developed for training and practice purposes. Planning and rehearsing disaster recovery procedures, using information gathered from real-life project risk experiences, can contribute enormous value and capacity to an organisation's risk management capability. In this regard, the benefits of learning from experiences of failure will often exceed those derived from stories of success.

Confidentiality and currency

In creating a risk register, due regard must be given to any internal requirements for confidentiality of information, and also to any laws relating to third party information disclosure in the country where the stakeholder organisation (or the project) is located. The physical security of the risk register and its data must also be considered.

The currency of a risk register depends upon how long its contents remain useful to the organisation. All such resources have a 'shelf life'. This may be in terms of relevance of the content, or in terms of the information management techniques used to create, maintain and access the risk register.

The relevance of the content of an organisational risk register will change over time. The organisation might decide to carry out its project activities differently, for example, replacing some technologies or resources with others. For various reasons it might opt to discontinue any future involvement with particular types of projects. If a risk register is to maintain a 'live' quality, it is as important to cull redundant information as it is to add new data.

Developments and changes in information management techniques mean that it might be necessary for an organisation to enlarge its information storage capacity, upgrade operating systems, or migrate data from one system to another.

Although care is needed in dealing with any of these changing situations, they provide opportunities for an organisation to consider the state of its risk register and to take steps to improve its efficacy.

9.12 REVIEWING RMS PERFORMANCE

One of the most difficult things to deal with in implementing and maintaining a risk management system in an organisation is evaluating its performance. After all, if risks are successfully managed so that either they do not eventuate or do not have the impact they might have had if unmanaged, how is the organisation to measure the invisible benefit? What are the key performance indicators and critical success factors for a risk management system in a project stakeholder organisation? Obviously these must be determined by each organisation and will be based upon the objectives established for the RMS as a whole and the functions of the constituent parts. However, difficulty will almost certainly still be encountered in attempting to measure system costs and value system benefits. Even if this can be accomplished, what benefit to cost ratio should be expected for the RMS, and how can it be benchmarked against the systems of other organisations? How frequently should RMS performance reviews be conducted?

The whole area of RMS performance is considerably under-researched to date; hence the multitude of questions in the paragraph above. Few answers are available in current risk management literature. Like any other management technique, the performance of a RMS *should* be reviewed. For the most part, a review will have to rely upon the critical assessment of feedback evidence from staff involved in using the RMS and much of this will be subjective opinion. Provided this is gathered in a methodical way, and validated wherever possible by triangulation and comparison, the results should be acceptable. Sourcing the data from a good cross-section of the organisation will help to achieve this. It will also minimise the risk of subversion or manipulation of the risk management system itself, as noted in the banking example earlier in this chapter.

Some clues for the RMS review process are described in table 9.3. The review focus is assumed to be on system performance from the perspective of the various constituent stages of a RMS. An organisation might choose initially to limit the performance review to one specific project, or one aspect of the system, on the grounds of simplifying and speeding up the review process. Eventually, however, reviews should embrace at least several (or a series) of projects and the whole RMS.

It will be seen from table 9.3 that most of the suggested performance criteria are based upon subjective assessment. Some guidance on RMS benchmarking can be obtained from the risk management maturity model presented at the beginning of this chapter, but for realistic comparison the co-operative collaboration of other organisations must be sought.

The frequency of RMS performance review should be at least annually for an organisation of level 2 risk management maturity. For a level 3 organisation the review could be conducted every two to three years, and at level 4 level this might be expanded to once every

Table 9.3 Performance review focus for a project stakeholder organisation RMS

RMS STAGE	PERFORMANCE FOCUS	SUGGESTED PERFORMANCE CRITERIA
Risk identification	Effectiveness of risk identification techniques and processes	What difficulties did staff experience in using techniques? What logistical problems were encountered in the identification process? How many foreseeable risks were missed and subsequently discovered later in the project? How many unforeseen problems were actually encountered later in the project? How realistic were the subjective assessments?
Risk analysis	Effectiveness of risk analysis techniques and processes	How accurate and reliable were any quantitative assessments? How effective were risk mitigation plans?
Risk response	Appropriateness and effectiveness of risk response decisions	How effective was risk transfer action? What comparisons can be made between before/after treatment risk severity scores or cluster maps (for a sequential series of projects)? Has the contingency spend rate per project decreased?
Risk monitoring and control	Effectiveness of risk monitoring and control procedures	Do any procedures overlap with other management actions (e.g. value management, quality management, safety management)? Entry rate for new entry material decreasing?
Risk recording and archiving	Adequacy and effectiveness of risk register	Is the risk severity of new entry material increasing or decreasing? Has the risk register yielded information of added value for case studies, disaster recovery plans and rehearsals, etc.?

four to five years. Five years is probably the maximum permissible interval between RMS performance reviews, since this roughly corresponds with the time normally required to plan, develop and implement technological and organisational change successfully in most industries. Where there is opportunity and capacity for reckless decision-making in the organisation, however, there is little alternative to adopting review procedures which are virtually continuous.

9.13 CHAPTER SUMMARY

The risk management maturity model should enable organisations to assess the extent to which they currently engage in formal risk management practices. It provides goals for improvement. Organisations need to establish a clear risk management policy and an implementation strategy for a RMS, as creating a formal RMS is a project in itself.

The organisation must decide upon the type of RMS (multiple project systems; dual project and internal; or seamless single system) to be adopted. Careful design of the system framework is needed. In the early stages of risk management maturity a simple spreadsheet approach may suffice. Issues of responsibility for the RMS must be resolved, and responsibility should be transparent, extending to the highest levels within the organisation. System building and performance will be enhanced if due attention is paid to communication, trialling techniques and training staff. The effectiveness of each part of the RMS should be tested, and appropriate periodic performance review mechanisms put in place.

While the general ground of building a RMS has been covered here, readers intending to undertake this in practice are recommended to further reading in the area of change management.

In the concluding chapter, another facet of risk – the management of opportunity – is considered.

CHAPTER 10

OPPORTUNITY MANAGEMENT

10.1 INTRODUCTION

Earlier in this book we noted that the prevailing view of risk – as a negative concept – is under challenge. Writers are now suggesting that focusing only on the consequences of risk events that may threaten or damage an organisation is too negative. It is thought that a persistently negative view of risk may lead to the abandonment of otherwise potentially successful projects; demotivation of project teams; and overly pessimistic attitudes on the part of organisations and the individuals within them.

Instead of concentrating solely on adverse risks, argue the proponents of the 'upside' risk view, an organisation should seek to exploit opportunities arising from its project involvement. Achieving balance between competing objectives is claimed as the essential aim of project risk management, and opportunity should be recognised as the natural obverse of a two-sided coin called 'uncertainty', which hitherto has only been proffered with its adverse risk face showing.

How appropriate is this dual perspective? How does it affect the processes of organisational risk management that we have already described? These are the issues addressed in this chapter.

10.2 THE 'ADVERSE IMPACT' RISK PERSPECTIVE

The concept of risks as events with outcomes that threaten the project objectives of a stakeholder organisation is well entrenched. Indeed, a negative view of risk is prevalent throughout society. It is reinforced in almost every book written about risk and risk management, or about projects and project management. This book is no exception. Even where writers are careful to make a connection between risk and opportunity, the treatment is usually brief and invariably the examples presented are slanted towards illustrating risks with adverse outcomes. Again, this book is no exception.

Why is this so, particularly since any good dictionary will offer definitions of risk in terms of both threat of loss and opportunity to profit? Yet earlier in this book, we too followed the trend and deliberately chose to adopt a definition of risk which emphasised the negative view. Was that an appropriate choice?

Several factors serve to explain the grip that the negative view of risk has on contemporary risk management.

In the first place, qualitative approaches to risk assessment have largely superceded purely quantitative techniques in modern risk management, mainly on the grounds of applicability and practicality for risk management purposes. We explored this in chapter 7, and noted that, for many risks involving project stakeholders, reliable data for quantitative risk assessment were simply not available; nor were they always needed. However, the mathematical underpinnings to quantitative risk assessment techniques are unbiased in terms of positive or negative risk impacts, and in any case concentrate almost exclusively upon the probability of occurrence of risk events. Qualitative risk assessment, on the other hand, relies heavily upon linguistic descriptors for alternative risk conditions in terms of likelihood, consequence and duration of exposure. These too were illustrated in chapter 7. By their nature, linguistic descriptors are often difficult to express in neutral terms. We can rarely use the same words to describe both horror and delight (although that does not deter contemporary youth from gleefully using negative terms to denote positive extremes: e.g. 'wicked!' and 'killer!'). The shift to more qualitative assessment has reinforced the adverse view of risk simply because the need for negative descriptors outweighed any demand for positives.

The second reason is linked to the first. In our experience, many people (and most project stakeholder organisations) tend towards risk aversion in their decision-making, especially in relation to follow-up decisions after an initial risk-seeking choice. We alluded to this in chapter 9 when we said that a person's risk attitude is not necessarily consistent over the whole range of decision-making. For example, research has shown that construction contractors tend to display risk-seeking (opportunistic?) behaviour when bidding for new work, but switch to risk aversion (threat protection?) after winning a tender. Similarly, someone deciding to sail solo around the world, or walk to the North Pole, has made an initial risk-seeking decision to exploit an opportunity. Almost every decision the adventurer makes after that, however, is made with the risk-averse intention of minimising threats to the achievement of his or her objective. Stakeholder organisations tend to follow the same pattern in projects of all kinds, although not all stakeholders are involved in the initial, opportunity-seeking, project decision.

In most cases, the organisational structure adopted for projects also precludes a more risk-seeking opportunistic approach to risk management. Parkin (1996) notes that much of project decision-making is guided by codes and specifications and that control of knowledge and information is the decision-making power source of professional groups or highly skilled technicians. He subscribes to two contrasting modes of organisational management: the administrative mode of ordering (structure, hierarchy, function, control, adaptability); and the enterprise mode of ordering (meaning, interests, power, politics, symbolism). The contrast between the two modes is in terms of interests, conflict and power. For interests, the administrative mode of organisational management will try to achieve a set of common objectives and work closely together to achieve them; while the enterprise mode recognises individual *and* group interests. The administrative mode of organising seeks harmony and the removal of conflict; while the enterprise mode accepts conflict as a natural, and often desirable, feature of organisations. In terms of power, the administrative organisation prefers clear lines of authority, leadership and control; while the enterprise mode recognises the power of groups and individuals in resolving conflicts of interest in the organisation.

Given the above characteristics, most projects would be identifiable as predominantly ordered under the administrative organisational management mode. Since the aims of the administrative mode are self-evidently intended to protect the organisation from threat, it is hardly surprising that project risk management focuses almost exclusively on the negative concept of risk. In fact, stakeholder organisations that happen to be enterprise-oriented may sometimes encounter problems with their involvement in administratively ordered project organisation structures. They experience difficulty in conforming to 'expected' norms and attract (often unfairly) reputations for being 'mavericks' or for failing to be 'team members'.

Finally, the influence of education and training should not be ignored. While any normal population should naturally include roughly equal proportions of risk-seeking (eager to exploit opportunities) and risk-averse (determined to avoid threats) people, nurture attempts to manipulate nature in terms of how we are educated and trained to handle projects. Increasingly, projects are carefully designed, engineered and managed by people with professional qualifications. Professional education, and the requirements for membership of professional institutions, are largely driven by criteria and content which are more analytical than creative, more focused upon administrative efficiency than entrepreneurship, more concerned with prudence than daring.

A negative perspective of risk in contemporary project risk management therefore simply reflects the reality of the sentient management environment of most projects. It is self-perpetuating through the attitudes, education, training, and roles of many of the participants. Before deciding if the negative perspective is more appropriate, however, the counter-perspective must be explored.

10.3 THE 'RISK OPPORTUNITY' PERSPECTIVE

Writers challenging the essentially negative view of project risk management argue that it emphasises the management of bad luck and fails to capitalise on good luck. Another argument states that the negative approach does not permit consideration of potential trade-offs between project objectives; that opportunities levered off one objective cannot be used to enhance achievement in another, simply because the management approach is focused exclusively upon potential threats to each objective.

It is also suggested that risk management is usually delayed until the project and its scope are sufficiently defined, because having more information available will permit risks to be dealt with more effectively. The delay thus limits any early systematic search for opportunities to exploit in terms of improving project performance.

The first argument is somewhat simplistically reductionist, since it assumes that every opportunity must have positive outcomes and that every threat will be completely negative. It fails to consider that exploiting opportunities will almost certainly introduce more risks and change the characteristics of others. It also fails to consider that treating the negative effects of threats may produce other positive results.

The argument, that potential trade-off between project objectives is neglected with a purely negative risk perspective ignores the modern development of value management techniques which are deliberately aimed at clarifying objectives at the inception of projects, so that the key project stakeholders can achieve a common understanding about them. An effective value management intervention in a project should include the explicit exploration of opportunistic trade-off between objectives.

Delaying the use of risk management on a project until sufficient information is available runs counter to the purpose of risk management and fails to appreciate the benefits of a dynamic RMS. Risk management is not a one-off intervention at a single point during a project, but a systematic and ongoing approach to dealing with risks over the whole life of the project. The project RMS should be implemented at the commencement of the project, so that it is able to influence the outcomes of decision-making at the earliest possible stage.

Although the arguments presented to date may be weak, the positive 'opportunity' view of risk and risk management is valid. The concept of 'risk' and 'reward' is well understood in financial investment management, and the notion of 'threat' and 'opportunity' is not far distant from this. The two may be seen as opposite extremes on a continuum of risk, as shown in figure 10.1.

Figure 10.1 Threats and opportunities in risk management

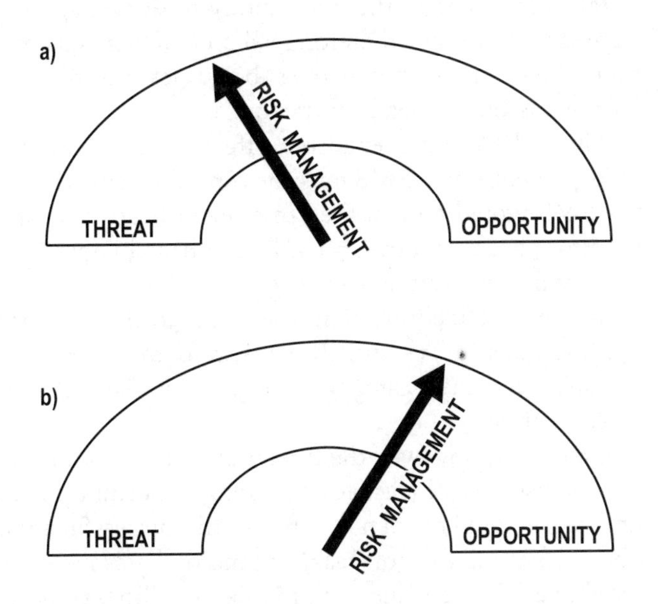

Given this continuum, two inferences can be made. The greater the focus on threats in risk management, the greater the number of opportunities that may be missed. Conversely, the greater the focus on opportunities, the greater the exposure to threats will be, since all opportunities bring with them their own 'downside' potential, and less effort will be directed towards dealing with existing threats.

10.4 IMPLICATIONS FOR RISK MANAGEMENT

Accepting a dual view of risk has some implications for project risk management.

In the first place, the consideration of opportunities should not be allowed to distract the organisation from properly assessing the nature and magnitude of the threats it faces. Surely it is better to miss an opportunity than to overlook a threat? In other words, threats to the achievement of project objectives should be accorded due priority in risk management.

Secondly, the people given the task of managing threats may not be best suited to explore and exploit opportunities. It should not be assumed that everyone involved with risk management will be equally adept in dealing with both extremes of the risk continuum. The increasing proliferation of specialist (subcontractor) project stakeholders is evidence of this. The inability of a single stakeholder to deliver all the elements of a project (tasks, technologies, resources, organisation) leads inevitably to the introduction of others who, because of their specialised knowledge or skills, are able and willing to take advantage of the opportunity to deliver specific project sub-elements. In a sense therefore, subcontracting is as much an opportunistic response to risk by the subcontractor as it is a means of transferring risk by the head contractor.

Given these arguments, the effect of incorporating opportunity management into a RMS must be carefully considered. The stages of the RMS would remain the same: identification; analysis; response; monitoring and control; recording and archiving, but some differences will occur within each stage.

In the risk identification stage, the techniques for recognising opportunities are essentially similar to those used for identifying threats, and the risk categories suggested in figure 2.5 would be applicable to either.

Effectively, however, the threats should be identified first, in order to establish a clear basis for exploring opportunities. Thus the question 'what could happen to threaten the successful attainment of this objective (in terms of task, technology, resource, organisation)?' would be followed (and not preceded) by 'how could this success be improved?'

Precise opportunity statements should flow from the identification stage: there is a chance 'p' that benefit 'b' will be obtained if opportunity 'c' is exploited during the period 't'. Note that this statement differs from that suggested in chapter 6 for adverse risk events. In opportunity management, the probability attaches not to the occurrence of an event leading to a consequence, but to the attainment of a particular outcome if a particular decision is made.

For the risk analysis stage, some of the qualitative indicator scales proposed in chapter 7 must be adapted for use in assessing opportunities.

The likelihood indicator scale shown in table 7.1 remains unchanged but, as noted above, it would now represent the chance that a particular benefit could be gained from the opportunity that has been recognised.

The negative impact factor scale descriptors from table 7.2 must be modified to reflect the positive nature of opportunity benefits. Table 10.1 shows how this might be done. The indicator scale representing the duration of exposure (to the risk event or its negative impact) shown in table 7.3 remains the same, but for opportunity management the scale would now reflect the period over which opportunity benefits can be sustained.

Table 10.1 Interval descriptors for risk opportunity impacts

Interval descriptor	Amplification
1. Insignificant	Marginal beneficial impact upon any or all of project time, cost or quality objectives
2. Minor	Perceptible beneficial impact upon any or all of project time, cost or quality objectives
3. Moderate	Significant beneficial impact upon some (but not all) project time, cost or quality objectives
4. Major	Major beneficial impact upon some (but not all) project time, cost or quality objectives, and significant beneficial impact upon the rest
5. Huge	Great beneficial impact upon all project time, cost and quality objectives

The risk threat severity scale of table 7.5 still applies, but would indicate the potential magnitude of an opportunity. Theoretically, given the application of the 5-point rating scales suggested in chapter 7, it should be possible to score both threats and opportunities during the risk analysis stage. Then any opportunity risk achieving a score substantially higher than its threat counterpart could be accorded priority in terms of treatment. The validity of such priority should be carefully tested and justified, however.

The labels used in chapter 8 to describe alternative risk treatment options in the risk response stage of the RMS are inappropriate to indicate how risk opportunities might be dealt with. Baccarini (2002) and Hillson (2001) have proposed contrasting risk opportunity labels. These are set out in table 10.2, showing the contrast between alternative treatments for adverse risk threats and those for risk opportunities. Clearly, an exact correspondence between treatment options at either end of the risk continuum is not possible. This fact supports our earlier caution that threat and opportunity aspects of risk management should not be attempted simultaneously, especially since

Table 10.2 Contrasting 'threat' and 'opportunity' response options in project risk managment

Risk 'threat' response option	AMPLIFICATION	Risk 'opportunity' response option	AMPLIFICATION
1. Avoid	Take another course of action that does not involve this risk	1. Exploit	Aggressively seek to obtain the maximum benefit from the opportunity
2. Transfer	Pass the risk on to another project stake holder	2. Share	Pass on the risk opportunity to another project stakeholder, or or come to a co-operative sharing arrangement for any benefit
3. Reduce	Mitigate one or more of the risk threat components and retain the residual risk	3. Enhance	Improve one or more of the risk opportunity components before exploiting or sharing it
4. Retain	Retain the whole risk without further treatment	4. Ignore	Do not take any action over the risk opportunity for this project

the risk response stage of a RMS directly involves a critical decision-making phase for the system.

The monitoring and control stage of a RMS could be adapted to encompass opportunity management, but the processes involved in this stage are likely to be substantially different to those needed for monitoring and controlling risk threats. The processes used to capture subsequent knowledge from risk opportunity management should be conducted separately from those used to record risk 'threat' experiences, since the nature of the information and the manner in which it will be used in the two situations are unlikely to share similarities.

For strategic approaches to risk management, the strategic risk management indicator described in table 8.1 can be applied to opportunity management without amendment.

10.5 CHAPTER SUMMARY

Contrary to some writers, we believe that threat management and opportunity management are not equally important in terms of their influence upon the success of projects.

The success of projects is determined by the achievement of the objectives established for them. Realistically, most projects are undertaken not with the aim of maximising the exploitation of all possible opportunities but in order to fulfil pre-defined functions. A project which delivers those functions can still be considered successful, even though a potentially better project which could have delivered more benefits might have been within reach if identifiable opportunities had been exploited. The priority of a project stakeholder organisation should be to first ensure the attainment of its objectives.

Our view is therefore that, while threat and opportunity can be seen as opposite poles on a risk continuum, the prior focus of risk management in any project stakeholder organisation should always be upon events (and their consequences) that threaten the achievement of objectives. If these objectives are not attained, the project cannot be successful. On the other hand, if project objectives are achieved, but could have been exceeded, the project cannot be said to be unsuccessful.

Note that this is not to say that opportunities must be ignored. It is simply a matter of setting appropriate management priorities in terms of available resources, and also of ensuring that the staff engaged in risk management activity are best suited for the tasks they are carrying out.

Despite concluding with a chapter which appears to contradict the premise upon which this book commenced – that risk is the chance that an adverse event will occur during a stated period of time – we hope that throughout this book a consistent and practically useful approach to risk management in project organisations has been presented and demonstrated.

As with any management tool or technique, project risk management cannot be applied passively, nor should it be regarded as static. An active approach to the application of risk management will yield positive benefits to any project organization. Maintaining an active approach, through a continuing cycle of learning and improvement, should contribute much towards ensuring that these benefits are sustained over the long-term life of the organisation.

REFERENCES

Akintoye, A., Hardcastle, C., Chinyio, E., Beck, M. and Asenova, D. (2001) *Standardised Framework for Risk Assessment and Management of Private Finance Initiative projects.* Report No.5, School of the Built and Natural Environment, Glasgow Caledonian University.

Allen, T.J. (1985) *Managing the Flow of Technology: Technology Transfer and the Dissemination of Technological Information within the R&D Organization*. MIT Press, Cambridge, MA.

Ansoff, H.I. (1965) *Corporate Strategy*. Penguin, New York.

Arvai, J.L., Gregory, R. and McDaniels, T.L. (2001) Testing a structured decision approach: value-focused thinking for deliberative risk communication. *Risk Analysis*, 21(6): 1065–1076.

AS/NZS 3931 (1998) *Risk Analysis of Technological Systems – Application Guide.* Standards Australia, Homebush, NSW.

AS/NZS 4360 (1999) *Risk Management.* Standards Australia, Homebush, NSW.

Awakul, P. and Ogunlana, S.O. (2002) The effect of attitudinal differences on interface conflicts in large scale construction projects: a case study. *Construction Management and Economics*, 20: 365–377.

Baccarini, D. (1996) The concept of project complexity – a review. *International Journal of Project Management*, 14(4), 210–214.

Baccarini, D. (2002) The positive side of risk management – managing opportunities. *Australian Project Manager*, 22(2) (June): 16–18.

Bennett, J. and Ormerod, R. (1984) Simulation applied to construction projects. *Journal of Construction Management and Economics*, 2(3): 225–263.

Berelson, B. and Steiner, G. (1964) *Human Behaviour*. Harcourt, Brace Jovanovich, New York.

Berlo, D.K. (1960) *The Process of Communication*. Holt, Rinehart and Winston, New York.

Birnie, J. (1993) A behavioural study using decision analysis of building cost prediction by chartered quantity surveyors. Unpublished DPhil. thesis, University of Ulster, Jordanstown, Northern Ireland.

Boje, D.M. (1995) Stories of the storytelling organization: a postmodern analysis of Disney as 'Tamara-Land'. *Academy of Management Journal*, 38(4): 997–1036.

Booth, A.E. (1981) The design of management information systems to handle uncertainty and complexity: a critical review of current practice. Unpublished MPhil. dissertation, North East London Polytechnic, London.

Booth, S.A. (1993) *Crisis Management Strategy – Competition and Change in Modern Enterprises.* Routledge, London.

Borys, B. and Jemison, D.B. (1989) Hybrid arrangements as strategic alliances: theoretical issues in organizational combinations. *Academy of Management Review*, 14(2): 234–249.

Bowen, D.E. and Schneider, B. (1988) Services marketing and management: implications for organizational behaviour. *Research in Organizational Behavior*, 10: 43–80.

Bowen, P.A. (1993) A communication-based approach to price modelling and price forecasting in the design phase of the traditional building procurement process in South Africa. Unpublished PhD thesis, University of Port Elizabeth, South Africa.

Bowen, P.A., Cattell, K.S., Pearl, R.G., Hall, K.A. and Edwards, P.J. (2000) Group dynamics in building procurement teams. *Proceedings: CIB W92 International Conference: Information and Communication in Construction Procurement*, pp. 291–311. Santiago, Chile.

Burns, T. and Stalker, G. (1961) *The Management of Innovation*. Tavistock Institute of Human Relations, London.

Butler, R., Davies, L., Pike, R. and Sharp, J. (1991) Strategic investment decision making: complexities, politics, and processes. *Journal of Management Studies*, 28(4): 395–415.

Cairns, G. and Beech, N. (1999) Use involvement in organisational decision making. *Management Decision Journal*, 37(1): 14–23.

Cleland, D.I. (1998) *The Project Management Institute: Project Management Handbook*. Jossey-Bass, San Francisco.

Cleland, D.I. and King, W.R. (1983) *Systems Analysis and Project Management*, 3rd edn. McGraw-Hill, New York.

Cohen, L.J. (1979) On the psychology of prediction: whose is the fallacy? *Cognition*, 7: 385–407.

Cohen, L.J. (1981) Can human irrationality be experimentally demonstrated? *The Behavioural and Brain Science Journal*, 4: 317–370.

Cooper, D. and Chapman, C. (1987) *Risk Analysis for Large Projects: Models, Methods and Cases.* John Wiley and Sons, Chichester, UK.

Crozier, M. (1964) *The Bureaucratic Phenomenon*, English translation. University of Chicago Press, Chicago.

Cyert, R. and March, J. (1963) *A Behavioural Theory of the Firm*. Prentice-Hall, Englewood Cliffs, New Jersey.

Dance, F.E.X. (1967) Toward a theory of human communication. In *Human Communication Theory*, ed. F.E.X. Dance, pp. 289–309. Holt, Rinehart and Winston, New York.

De Brentani, U. (1989) Success and failure in new industrial services. *Journal of Product Innovation Management*, 6: 239–258.

Donaldson, T. and Preston, L.E. (1995) The stakeholder theory of the corporation: concepts, evidence, and implications. *Academy of Management Review*, 20(1): 65–91.

Dowling, J. and Pfeffer, J. (1975) Organizational legitimacy: social values and organizational behavior. *Pacific Sociological Review*, 18(1): 122–134.

Dulaimi, M.F., Ling, F.Y.Y. and Bajracharya, A. (2003) Organizational motivation and inter-organizational interaction in construction innovation in Singapore. *Construction Management and Economics*, 21: 307–318.

Dutton, J.E. (1986) The processing of crisis and non-crisis strategic issues. *Journal of Management Studies*, 23(5): 501–517.

Feldberg, M. (1975) *Organizational Behaviour: Text and Cases*. Juta and Company, Cape Town.

Fellows, R.F., Langford, D.A., Newcombe, R. and Urry, S.A. (1983) *Construction Management in Practice*. Construction Press, London.

Fisher, B.A. (1978) *Perspectives on Human Communication*. Macmillan, New York.

Fotheringham, W.C. (1966) *Perspectives on Persuasion*. Allyn and Bacon, Boston.

Freeman, R.E. (1984) *Strategic Management: A Stakeholder Approach*. Pitman, Boston.

Friedman, A.L. and Miles, S. (2002) Developing stakeholder theory. *Journal of Management Studies*, 39(1): 1-21.

Frooman, J. (1999) Stakeholder influence strategies. *Academy of Management Review*, 24(2): 191–225.

Gerbner, G. (1956) Towards a general model of communication. *Audio-Visual Communication Review*, 3: 3–11.

Griffith, A. (1985) *Buildability: The Effect of Design and Management of Construction*. Report, Department of Building, Heriot-Watt University, Edinburgh.

Gibson, K. (2000) The moral basis of stakeholder theory. *Journal of Business Ethics*, 26(3): 245–257.

Hambrick, D.C. and Mason, P.A. (1984) Upper echelons: the organisation as a reflection of its top manager. *Academy of Management Review*, 27: 271–291.

Handy, C. (1985) *Understanding organizations*, 3rd edn. Penguin, London.

Harrison, J.S. and St. John, C.H. (1996) *Strategic Management of Organisation and Stakeholders*. South-Western College Publishing, Cincinati, Ohio.

Heath, R.L. and Abel, D.D. (1996) Types of knowledge as predictors of company support: the role of information in risk communication. *Journal of Public Relations Research*, 8: 35–55.

Heath, R.L., Seshadri, S. and Lee, J. (1998) Risk communication: a two-way analysis of proximity, dread, trust, involvement, uncertainty, openness/accessibility, and knowledge on support/opposition toward chemical companies. *Journal of Public Relations Research*, 10(1): 35–56.

Hertz, D.B. and Thomas, H. (1984) *Practical Risk Analysis: An Approach through Case Histories*. John Wiley and Sons, Chichester.

Hillson, D. (2001) Extending the risk process to manage opportunities. *Proceedings: Fourth European Project Management Conference*. PMI, London, 6–7 June. www.risksig.com

Hillson, D. (2002) *Risk Management Maturity Level Development*. Risk Management Specific Interest Group. Project Management Institute, Newtown Square, Pennsylvania.

Jawahar, I. and McLaughlin, G.L. (2001) Toward a descriptive stakeholder theory: an organizational life cycle approach. *Academy of Management Review*, 26(3): 397–414.

Johnson, W. (1951) The spoken word and the great unsaid. *Quarterly Journal of Speech*, 32: 421.

Johnson, G.J. and Scholes, K. (1984) *Exploring Corporate Strategy*. Prentice-Hall, Englewood Cliffs, New Jersey.

Kahneman, D. and Tversky, A. (1979) Prospect theory: an analysis of decision under risk. *Econometrica*, 47: 263–291.

Kahneman, D., Slovac, P. and Tversky, A. (1982) *Judgement under Uncertainty: Heuristics and Biases*. Cambridge University Press, Cambridge.

Katz, D. and Kahn, R.L. (1978) *The Social Psychology of Organizations*, 2nd edn. Wiley, New York.

Knight, K. (1976) Matrix organization: a review. *Journal of Management Studies*, May: 111–130.

Knoke, D. (2000) *Changing Organizations*. Westview Press, Boulder, Colorado.

Kotler, J. and Heskett, J. (1992) *Corporate Culture and Performance*. Free Press, New York.

Lasswell, H.D. (1948) The structure and function of communication in society. In *The Communication of Ideas*, ed. L. Bryson, pp. 37–51. Institute of Religious and Social Studies, New York.

Lawrence, P.R. and Lorsch, J.W. (1967) *Organisation and Environment: Managing Differentiation and Integration*. Harvard University Press, Massachusetts.

Li, Bing. (2003). Risk management of construction public private partnership projects. Unpublished PhD thesis, School of the Built and Natural Environment, Glasgow Caledonian University.

Lievens, A. and Moenaert, R.K. (2000) Project team communication in financial service innovation. *Journal of Management Studies*, 37(5): 733–766.

Loosemore, M. (1996) Reactive crisis management in construction projects: a longitudinal investigation of communication and behaviour patterns with a grounded-theory framework. Unpublished PhD thesis, University of Reading, Reading.

Mak, S.W. (1992) Risk management in construction: a study of subjective judgements. Unpublished PhD thesis, University College, London.

Maslow, A.H. (1943) A theory of human motivation. *Psychological Review*, 50: 370–396.

Mayo, E. (1933) *The Human Problems of an Industrial Civilization*. Macmillan, New York

Mayo, E. (1949) Hawthorne and the Western Electric company. In *Organisation Theory – Selected Readings*, 2nd edn (1984), ed. D.S. Pugh, pp. 279–292. Penguin Books, Harmondsworth.

McCann, D. and Margerison, C. (1989) High performance teams. *Training and Development Journal*, 43: 52–60.
McGregor, D. (1960) *The Human Side of Enterprise*. McGraw-Hill, New York.
Mearns, K., Flin, R. and O'Connor, P. (2001) Sharing 'worlds of risk'; improving communication with crew resource management. *Journal of Risk Research*, 4(4): 377–392.
Mills, A.E. (1967) *The Dynamics of Management Control Systems*. London Business Publications, London.
Mintzberg, H. (1979) *The Structuring of Organisations*. Prentice-Hall, Englewood Cliffs, New Jersey.
Mintzberg, H. (1983) *Structures in Fives: Designing Effective Organisations*. Prentice-Hall, Englewood Cliffs, New Jersey.
Murray, M. (1980) The interaction of the theory of corporate planning and the microeconomic theory of the firm in the development of corporate planning models. Unpublished MSc. dissertation, University of Bath, Bath.
Newcomb, T.M. (1953) An approach to the study of communicative acts. *Psychological Review*, 60: 393–404.
Newman, E.B. (1948) Hearing. In *Foundations of Psychology*, eds. Boring, E.G, Langfeld, H.S. and Weld, H.A., pp. 313–350. John Wiley and Sons, New York.
Otway, H. (1992) Public wisdom, expert fallibility: toward a contextual theory of risk. In *Social Theories of Risk*, eds Krimsky, S. and Golding, D., pp. 215–228. Praeger, Westport, CT.
OED (1989) *Oxford English Dictionary*. Oxford University Press.
Palenchar, M.J. and Heath, R.L. (2002) Another part of the risk communication model: analysis of communication process and message content. *Journal of Public Relations Research*, 14(2): 127–158.
Parkin, J. (1996) *Management Decisions for Engineers*. Thomas Telford, London.
Post, J.E., Frederick, W.C., Lawrence, A.T., and Weber, J. (1996) *Business and Society: Corporate strategy, Public Policy and Ethics*. McGraw-Hill, Sydney.
Preece, C.N., Moodley, K., and Smith, A.L. (1998) *Corporate Communications in Construction: Public Relations Strategies for Successful Business and Projects*. Blackwell Science, Oxford.
Rajogopolan, N., Raheed, A.M. and Datta, D.K. (1993) Strategic decision processes: critical review and future directions. *Journal of Management*, 12(2): 349–284.
Ross, R.S. (1965) *Speech Communication: Fundamentals and Practice*. Prentice-Hall, Englewood Cliffs, New Jersey.
Rowley, T.J. (1997) Moving beyond dyadic ties: a network theory of stakeholder influences. *Academy of Management Review*, 22: 887–910.
Rowlinson, S. (2001) Matrix organizational structure, culture and commitment: a Hong Kong public sector case study of change. *Construction Management and Economics*, 19: 669–673.
Rowlinson, S. and Root, D. (1996) *The Impact of Culture on Project Management*. Final Report to the British Council, Hong Kong. Department of Real Estate and Construction, Hong Kong University.
Royal Society (1991) *Report of the Study Group on Risk: Analysis, Perception, Management* (Chairman: Professor Sir Frederick Warner). The Royal Society, London.
Sagan, S.D. (1994) Toward a political theory of organisational reliability. *Journal of Contingencies and Crisis Management*, 2(4): 228–240.
Schachter, S. (1951) Deviation, rejection and communication. *Journal of Abnormal and Social Psychology*, 46: 190–207.
Schramm, W. (1955) *The Process and Effects of Mass Communication*. University of Illinois Press, Urbana.

Scott, S.G. and Lane, V.R. (2000) A stakeholder approach to organisational identity. *Academy of Management Review*, 25(1): 43–62.

Selznick, P. (1957) *Leadership in Administration*. Row Peterson, Evanston, Illinois.

Shannon, C. and Weaver, W. (1949) *The Mathematical Theory of Communication*. University of Illinois Press, Urbana.

Silverman, D. (1970) *The Theory of Organisations*. Heinemann, London.

Simon, H.A. (1948) *Administrative Behaviour*. Macmillan, New York.

Simon, H.A. (1960) *The New Science of Management Decision*. Harper and Row, New York.

Simon, H.A. (1969) *The Sciences of the Artificial*. MIT Press, Cambridge, Massachusetts.

Simon, H.A. (1973) Decision making and organizational design. In *Organisation Theory: Selected Readings*, ed. P.S. Pugh, pp. 202–223. Penguin, Middlesex, England.

Simons, R.H. and Thompson, B.M. (1998) Strategic determinants: the context of managerial decision making. *Journal of Managerial Psychology*, 13(1/2): 7–21.

Skitmore, R.M. (1986) A model of the construction project selection and bidding decision. Unpublished PhD thesis, Department of Civil Engineering, University of Salford, Salford.

Slovic, P. (1972) Psychological study of human judgment: implications for investment decision-making. *Journal of Finance*, 27: 779–799.

Smith, N.J. (1999) *Managing Risk in Construction Projects*. Blackwell Science Ltd, Oxford.

Snary, C. (2002) Risk communication and the waste-to-energy incinerator environmental impact assessment process: a UK case study of public involvement. *Journal of Environmental Planning and Management*, 45(2): 267–283.

Stopford, J.M. and Wells, L.T. (1972) *Managing the Multi National Enterprise: Organisation of the Firm and Ownership of Subsidiaries*. Basic Books, New York.

Thomsen, J.D. (1967) *Organisations in Action*. McGraw-Hill, New York.

Thompson, K.M. and Bloom, D.L. (2000) Communication of risk assessment information to risk managers. *Journal of Risk Research*, 3(4): 333–352.

Thwaites, D. (1992) Organizational influences on the new product development process in financial services. *Journal of Product Innovation Management*, 9: 303–313.

Tubbs, S.L. and Moss, S. (1981) *Interpersonal Communication*. Random House, New York.

Tushman, M.L. (1979) Work characteristics and subunit communication structure: a contingency analysis. *Administrative Science Quarterly*, 24: 82–97.

Von Bertalanffy, L. (1969) *General Systems Theory: Essays on its Foundation and Development*. Braziller, New York.

Von Neumann, J. and Morganstern, O. (1944) *The Theory of Games and Economic* Behaviour (3rd edn, 1953). Princeton University Press, Princeton, New Jersey.

Wagner, H.M. (1971) *Operations Research*. Prentice-Hall, Englewood Cliffs, New Jersey.

Westley, B.H. and MacLean, M.S. Jr. (1957) A conceptual model for communications research. *Journalism Quarterly*, 34: 31–38.

Williams, T.M. (1999) The need for new paradigms for complex projects. *International Journal of Project Management*, 17(5): 269–273.

Winch, G. (1998) Zephyrs of creative destruction: understanding the management of innovation in construction. *Building Research and Practice*, 26(4): 268–279.

Woodward, J. (1965) *Industrial Organization*. Oxford University Press, London.

INDEX